Carotenoids

Carotenoids

Special Issue Editor

Volker Böhm

MDPI • Basel • Beijing • Wuhan • Barcelona • Belgrade

MDPI

Special Issue Editor
Volker Böhm
Friedrich Schiller University Jena
Germany

Editorial Office
MDPI
St. Alban-Anlage 66
4052 Basel, Switzerland

This is a reprint of articles from the Special Issue published online in the open access journal *Antioxidants* (ISSN 2076-3921) from 2018 to 2019 (available at: https://www.mdpi.com/journal/antioxidants/special_issues/carotenoids_health).

For citation purposes, cite each article independently as indicated on the article page online and as indicated below:

LastName, A.A.; LastName, B.B.; LastName, C.C. Article Title. *Journal Name* **Year**, *Article Number*, Page Range.

ISBN 978-3-03921-864-6 (Pbk)
ISBN 978-3-03921-865-3 (PDF)

Cover image courtesy of Angelika Böhm.

Contents

About the Special Issue Editor

Volker Böhm studied Food Chemistry at Münster University (Germany). He then completed his Ph.D. studies on the analysis of residues of coplanar PCB congeners in food and human milk and was awarded his Ph.D. in 1992 from Münster University. In 1999, he was appointed as Lecturer at Jena University (Germany) covering the topic Antioxidative Carotenoids and Polyphenols—Analysis, Contents in Food, and Intestinal Absorption. He was promoted to Adjunct Professor (Food Chemistry) in 2019.

Dr. Böhm began as Senior Scientific Assistant at Jena University in 1999 and progressed to Research Group Leader, a position he has held since 2005. He was responsible for coordinating the large EU-funded project LYCOCARD (2006–2011). His main research topics are secondary plant products with a focus on carotenoids and polyphenols. His recent investigations have examined the contents of carotenoids and polyphenols, their behavior during food processing, as well as their biological activities and intestinal absorption.

Preface to "Carotenoids"

Carotenoids are a group of natural pigments consisting of more than 750 compounds known so far. They are mostly yellow, orange, or red in color, due to the system of conjugated double bonds. This structural element is also responsible for the good antioxidant properties of many carotenoids. Carotenoids have shown numerous biological activities (not only as provitamin A), making them an interesting topic for researchers of various disciplines all over the world looking for, e.g., preventive properties of fruits and vegetables. As lipophilic compounds, their uptake and storage in the body are dependent on various conditions. In vitro and in vivo data showed stimulating and inhibitory effects of matrix compounds on bioaccessibility and bioavailability of carotenoids.

This Special Issue includes 11 peer-reviewed papers, including eight original research papers and three reviews. Together, they represent the most recent advances in carotenoid research, in addition to the search for antioxidant properties. Chapters include overviews on the photoprotective properties of carotenoids as well as on the activities of carotenoids related to liver health. The research papers present data on the effect of degree of ripeness on carotenoids pattern in rosehip and possibilities to use shrimp waste as a source of carotenoids. Other investigations characterized apocarotenoids in microalgae and the properties of inclusion complexes of lycopene and beta-cyclodextrin. Biological activities of synthesized retinoyl-flavonolignan hybrids were also reported. In addition, the effects of in vitro digestion of human milk on the micellization of carotenoids were investigated.

All authors are acknowledged for their valuable contributions. I would also like to acknowledge all reviewers for their constructive suggestions. Special thanks are to the publishing team of the journal *Antioxidants* for all their help in compiling this Special Issue.

Volker Böhm
Special Issue Editor

antioxidants

MDPI

Editorial

Carotenoids

Volker Böhm

Institute of Nutritional Sciences, Friedrich Schiller University, Dornburger Str. 25-29, 07743 Jena, Germany;
volker.boehm@uni-jena.de

Received: 22 October 2019; Accepted: 26 October 2019; Published: 29 October 2019

Carotenoids are a group of natural pigments, consisting of more than 750 compounds known so far. Their colours are mostly yellow, orange, or red due to the system of conjugated double bonds. This structural element is also responsible for the good antioxidant properties of many carotenoids. Carotenoids have shown numerous biological activities (not only as provitamin A), making them an interesting topic for researchers of various disciplines all over the world looking for, e.g., preventive properties of fruits and vegetables. As lipophilic compounds, their uptake and storage in the body are dependent on various conditions. In vitro and in vivo data showed stimulating and inhibitory effects of matrix compounds on bioaccessibility and bioavailability of carotenoids [1].

This special issue highlights some of the recent advances in carotenoid research, showing on the one hand the status quo and giving on the other hand new insight in functions and physiological relevance. Al-Yafeai et al. investigated rosehips of *Rosa rugosa* at different degrees of ripeness and showed that maturity stage significantly affected contents of bioactive ingredients as well as antioxidant capacity. Thus, harvesting date can be chosen depending on the contents of bioactive molecules. In addition, the authors fully characterized *(5′Z)*-rubixanthin (gazaniaxanthin) as the main (Z)-isomer of rubixanthin in hips of *R. rugosa* by using HPLC-MS/MS and NMR [2]. Chitong et al. investigated biological activities of astaxanthin extracted from shrimp waste. They discussed a possible use of this extract in dietary supplements for skin health applications [3]. Apocarotenoids are cleavage products of carotenoids. These molecules often showed biological activity. Thus, Zoccali et al. extracted four microalgae strains and characterized the apocarotenoids therein [4]. Sandmann reviews structure dependence of antioxidant protection from UV and light stress by carotenoids. Substitutions at the hydrocarbon molecule are important for good singlet oxygen quenching and radical scavenging [5].

The review by Elvira-Torales et al. gives an overview on beneficial health-related effects of carotenoids with special emphasis on liver health. The authors described, e.g., how carotenoids can interact in patients with non-alcoholic fatty liver disease (NAFLD) [6]. Results of an experiment using Wistar rats are presented by Róvera Costa et al. The authors evaluated the liver protective effects of supplementation (10 weeks) with lycopene on non-alcoholic fatty liver disease [7].

Chambers et al. synthesized special retinoyl-flavonolignan hybrids and investigated their antioxidant properties [8]. The paper of Karpiński and Adamczak reports results of antimicrobial activity of the carotenoid fucoxanthin against 20 bacterial species [9]. Balić et al. review the photoprotective properties of carotenoids in skin. Dietary carotenoids accumulate in the epidermis and act as a protective barrier to various environmental influences [10].

In vitro digestion experiments with human milk were done by Xavier et al. These authors investigated how the lactation stage (colostrum, mature milk) affected the carotenoid contents in micelles [11]. Wang et al. prepared inclusion complexes of lycopene in β-cyclodextrin and characterized these complexes by looking for stability and antioxidant activity. Lycopene was embedded into the cavity of β-cyclodextrin with a 1:1 stoichiometry [12].

Thus, this special issue presents a lot of various results, highlighting preventive effects of carotenoids in diverse conditions. In addition, some properties of single carotenoids are presented to deepen the knowledge on this fascinating group of compounds.

Conflicts of Interest: The author declares no conflict of interest.

References

1. Westphal, A.; Böhm, V. Carotenoids. Properties, distribution, bioavailability, metabolism and health effects. *Ernahr. Umsch.* **2015**, *62*, 196–207.
2. Al-Yafeai, A.; Bellstedt, P.; Böhm, V. Bioactive compounds and antioxidant capacity of *Rosa rugosa* depending on degree of ripeness. *Antioxidants* **2018**, *7*, 134. [CrossRef] [PubMed]
3. Chintong, S.; Phatvej, W.; Rerk-Am, U.; Waiprib, Y.; Klaypradit, W. In vitro antioxidant, antityrosinase, and cytotoxic activities of astaxanthin from shrimp waste. *Antioxidants* **2019**, *8*, 128. [CrossRef] [PubMed]
4. Zoccali, M.; Giuffrida, D.; Salafia, F.; Socaciu, C.; Skjånes, K.; Dugo, P.; Mondello, L. First apocarotenoids profiling of four microalgae strains. *Antioxidants* **2019**, *8*, 209. [CrossRef] [PubMed]
5. Sandmann, G. Antioxidant protection from UV- and light-stress related to carotenoid structures. *Antioxidants* **2019**, *8*, 219. [CrossRef] [PubMed]
6. Elvira-Torales, L.; García-Alonso, J.; Periago-Castón, M. Nutritional importance of carotenoids and their effect on liver health: A review. *Antioxidants* **2019**, *8*, 229. [CrossRef]
7. Róvero Costa, M.; Leite Garcia, J.; Cristina Vágula de Almeida Silva, C.; Junio Togneri Ferron, A.; Valentini Francisqueti-Ferron, F.; Kurokawa Hasimoto, F.; Schmitt Gregolin, C.; Henrique Salomé de Campos, D.; Roberto de Andrade, C.; dos Anjos Ferreira, A.; et al. Lycopene modulates pathophysiological processes of non-alcoholic fatty liver disease in obese rats. *Antioxidants* **2019**, *8*, 276. [CrossRef] [PubMed]
8. Chambers, C.; Biedermann, D.; Valentová, K.; Petrásková, L.; Viktorová, J.; Kuzma, M.; Křen, V. Preparation of retinoyl-flavonolignan hybrids and their antioxidant properties. *Antioxidants* **2019**, *8*, 236. [CrossRef]
9. Karpiński, T.; Adamczak, A. Fucoxanthin-An antibacterial carotenoid. *Antioxidants* **2019**, *8*, 239. [CrossRef] [PubMed]
10. Balić, A.; Mokos, M. Do we utilize our knowledge of the skin protective effects of carotenoids enough? *Antioxidants* **2019**, *8*, 259. [CrossRef] [PubMed]
11. Xavier, A.; Garrido-López, J.; Aguayo-Maldonado, J.; Garrido-Fernández, J.; Fontecha, J.; Pérez-Gálvez, A. In vitro digestion of human milk: Influence of the lactation stage on the micellar carotenoids content. *Antioxidants* **2019**, *8*, 291. [CrossRef] [PubMed]
12. Wang, H.; Wang, S.; Zhu, H.; Wang, S.; Xing, J. Inclusion complexes of lycopene and β-cyclodextrin: Preparation, characterization, stability and antioxidant activity. *Antioxidants* **2019**, *8*, 314. [CrossRef] [PubMed]

![antioxidants logo] *antioxidants*

MDPI

Article

Bioactive Compounds and Antioxidant Capacity of *Rosa rugosa* Depending on Degree of Ripeness

Ahlam Al-Yafeai [1,2], Peter Bellstedt [3] and Volker Böhm [1,*]

[1] Institute of Nutritional Sciences, Friedrich Schiller University Jena, Dornburger Straße 25-29, 07743 Jena, Germany; Ahlam.Al-Yafeai@uni-jena.de
[2] Department of Biology, Science Faculty, Ibb University, Ibb, Yemen
[3] Institute of Organic Chemistry and Macromolecular Chemistry, Friedrich Schiller University Jena, Humboldtstraße 10, 07743 Jena, Germany; peter.bellstedt@uni-jena.de
* Correspondence: Volker.Boehm@uni-jena.de; Tel.: +49-364-194-9633

Received: 28 August 2018; Accepted: 25 September 2018; Published: 3 October 2018

Abstract: Maturity stage affects the bioactive compounds as well as the antioxidant capacity in the fruit. This study was designed to identify and quantify carotenoids, as well as to evaluate vitamin E, vitamin C, antioxidant capacity and total phenolic compounds of *Rosa rugosa* hips at different degrees of ripeness. HPLC (high performance liquid chromatography) analysis showed different types of carotenoids at different stages of maturity of *R. rugosa* hips with significant differences ($p < 0.05$), where the maximum concentration was observed at late harvesting. In the hips investigated, only α-tocopherol was detected, the maximum concentration of both vitamin E and vitamin C was obtained in the orange hips with significant difference ($p < 0.05$). On the other hand, the highest hydrophilic and lipophilic TEAC (Trolox equivalent antioxidant capacity) values, as well as total phenolic contents, were determined in the mature hips (red colour) with significant difference ($p < 0.0001$) and ($p < 0.001$) respectively, whereas ORAC (oxygen radical absorbance capacity) showed lower activity in the mature hips with significant difference ($p < 0.05$). Late harvesting is recommended if a high content of carotenoids is desired, while harvesting should be carried out earlier if a higher vitamin E and vitamin C content is desired, which in turn affects the antioxidants capacity.

Keywords: gazaniaxanthin; NMR (nuclear magnetic resonance); L-TEAC

1. Introduction

Recent developments in the fields of health and food have led to a renewed interest in natural compounds with antioxidant potential. A diet rich in antioxidant components has potential effects on the human health by reducing the risk of various diseases, for example cardiovascular diseases, cancers and age-related macular degeneration [1]. The selection of species/varieties with high contents of bioactive compounds and harvesting at the optimum time can promote the increase in the uptake of bioactive compounds from the fruits and vegetables. The genus Rosa contains over 100 species that are widely distributed mostly in Europe, Asia, the Middle East and North America [2]. *R. rugosa* hips resemble tomatoes with a 2–3 cm diameter and often a shorter height than their diameter. During maturation, the hips' colour begins to change on their sunny side and become completely coloured over time [3,4]. In the same vein, rosehips have a fairly long maturation period, harvesting once a year in the autumn after the first night frost. In recent decades, the rosehip fruit has been increasingly studied for its preventive properties [5]. Previous studies reported that rosehips contain higher amounts of various bioactive compounds than several other fruits such as: carotenoids, vitamins (particularly vitamins C, E and provitamin A), flavonoids, tannins and fatty acids [3,6,7]. Moreover, Al-Yafeai et al. (2018) [3] have reported that rosehips extract contained four (Z)-lycopene isomers. Furthermore, rosehips extract was able to scavenge reactive oxygen species (ROS) [8]. The *Rosa multiflora* hips have been traditionally utilised as herbal remedies and nutritional supplements for diseases treatment, including osteoarthritis, rheumatoid arthritis, chronic pain as well as in the cold and inflammation diseases [9]. Although *Rosa rugosa* is known to produce the most abundant and best tasting hips, most food products are based on the hips of *R. canina*, although they are small compared with the *R. rugosa* hips [10]. In addition, *Rosa canina* hips have been largely utilised in traditional folk medicine. Although the carotenoids composition as well as vitamin E contents in mature *R. rugosa* have been previously described [3], there is no work presenting how ripening stages influence *R. rugosa* hips composition as well as antioxidant capacity (AOC). This study aimed to address the following research question: effect of ripening times on the carotenoid, vitamin E and vitamin C contents as well as on antioxidant capacity in *R. rugosa* hips and how to improve levels of these bioactive compounds by choice of harvesting time. On the other hand, the present study was also designed to identify and quantify the type of (Z)-isomers of lycopene.

2. Materials and Methods

2.1. Chemicals

All chemicals and buffer salts were of analytical quality and the solvents for chromatography were of HPLC grade. HPLC grade water was produced using a MicroPure instrument (Thermo Electron LED GmbH, Niederelbert, Germany). Moreover, the chemicals were of the highest quality available (95–99%) and were used without purification. The standards of carotenoids (97–99%) were from CaroteNature (Münsingen, Switzerland), (all-E)-rubixanthin was supplied by DSM, Kaiseraugst, Switzerland. Pure tocotrienols (>97%) and tocopherols (>95%) were purchased from Davos Life Sciences (Singapore, Singapore) and Calbiochem (Darmstadt, Germany), respectively. DL-α-tocopheryl acetate (>96%), 2,2′-azinobis-(3-ethylbenzothiazoline-6-sulphonic acid) diammonium salt (ABTS), 2,2′-azobis-(2-amidinopropane) hydrochloride (AAPH), phosphate buffered saline (PBS; pH 7.4, 75 mM), 6-hydroxy-2,5,7,8-tetramethylchroman-2-carboxylic acid (Trolox), HCl and Folin-Ciocalteu phenol reagent (FCR) were obtained from Sigma-Aldrich (Taufkirchen, Germany). Thiourea, meta-phosphoric acid and copper (II) sulphate (CuSO4) were purchased from Merck (Darmstadt, Germany). L (+)-ascorbic acid (AA), fluorescein and 3, 4, 5-trihydroxybenzoic acid (gallic acid) were purchased from Fluka (Buchs, Switzerland). Na_2CO_3, NaCl, NaOH, $CH_3COONa \cdot H_2O$, $Na_2HPO_4 \cdot 2H_2O$, $NaH_2PO_4 \cdot 2H_2O$ and Trichloroacetic acid (TCA) were obtained from VWR (Darmstadt, Germany).

2.2. Samples Description

One kg of *Rosa rugosa* Thunb. hips were harvested at Dorndorf-Steudnitz 51°00′ N, 11°68′ E at different ripening stages from different bushes of *Rosa rugosa* in late summer to early autumn of 2016. The average temperature was between 26 °C and 17 °C and the hips were picked by hand once and then separated into three groups depending on the degree of ripeness (green, orange and red). For analysis, the hips were cleaned, first rinsed thoroughly for 1–2 min with tap water, then 1 min with distilled water. Before homogenization, the seeds, calyx and stem were removed, then samples were ground for 20 s with a mill (Retsch Grindomix GM 200, Haan, Germany) and stored at −25 °C until analysis.

2.3. Carotenoids and Vitamin E Determinations

2.3.1. Extraction Procedure

Extraction of carotenoids and vitamin E from raw *R. rugosa* hips at different ripening times was performed by methods previously described by Al-Yafeai et al. (2018) [3]. 35 mL of MeOH/THF (1/1, *v/v*), 200 mg magnesium oxide and 200 mg sodium sulphate were added to 500 mg of *R. rugosa* hips. The homogenization was carried out for 5 min by using ultra turrax, the supernatant was filtered and collected while the residue was extracted again, at least two times. The collected supernatants were concentrated in a rotary evaporator and the final volume was 5 mL. Saponification is a necessary step to determine xanthophyll contents in rosehips, one advantage of the saponification is that it avoids the problem of unwanted lipids and other interfering substances and to hydrolyse carotenoid esters. Saponification was conducted by using 10% methanolic KOH for 60 min at room temperature. All procedures were carried out under dim light to sidestep the photo-degradation. HPLC analysis of carotenoids, lycopene isomers and vitamin E was realised immediately after extraction.

2.3.2. Carotenoids Identification and Quantification

HPLC analysis of carotenoids was performed by using reversed-phase HPLC with a photo diode array (PDA) detector according to the method previously described by Al-Yafeai et al. (2018) [3]. The method was reproducible and specific, the identifications were achieved by comparison with external standards, or compounds were tentatively identified by comparison of retention times and DAD absorption spectra and mass spectra. On the other hand, the quantifications were done using a 5-point calibration curve ($r^2 \geq 0.999$) of external standards. The limits of detection (LOD) and limits of quantification (LOQ) of the carotenoids were calculated using the baseline noise signals: LOD was supposed to a signal-to-noise (S/N) ratio of 3:1, LOQ to that one of 10:1.

2.3.3. Analysis of Lycopene Composition

To increase the reliability of measures, chromatographic separation of the lycopene isomers was achieved using the isocratic C30-HPLC method according to the method reported by Al-Yafeai et al. (2018) [3]. The identification was carried out by comparing their retention times with those of external standards, while the quantification was achieved by 3-point calibration curve ($r^2 \geq 0.999$) of external standards of *(all-E)*-lycopene, *(13Z)*-lycopene, *(9Z)*-lycopene and *(5Z)*-lycopene by using specific extinction coefficients [11].

2.3.4. Isomerisation and NMR Analysis

As mentioned previously, *(5'Z)*-rubixanthin (gazaniaxanthin) is the main *(Z)*-isomer of *(all-E)*-rubixanthin as well as is one of the important features that characterised *R. rugosa* compared to *R. canina*. The major problem in isolating *(5'Z)*-rubixanthin using the fractionation method [3] is that the concentration was too low. In addition, invisible compounds having a molecular weight of more than 1000 *m/z* were simultaneously eluted. An alternative method for confirming our results was using isomerisation of *(all-E)*-rubixanthin and checking the isomers by using NMR according to a method reported by Arpin & Liannen-Jensen (1969) [12] and Kishimoto et al. (2005) [13]. Approximately 1 mg *(all-E)*-rubixanthin was dissolved in CDCl$_3$ and immediately subjected to NMR analysis. Special attention was made to circumvent exposure to sun-/daylight. All NMR experiments were performed on a 600 MHz Bruker Avance III spectrometer equipped with a 5 mm triple resonance cryogenic probe. Sample temperature was set to 293 K. The referencing was carried out using the residual 1H solvent peak of CDCl$_3$. The assignment was based on literature data [13] as well as predicted spectra (ACD Labs NMR Workbook 2017.2.1) and was confirmed by standard NMR experiments (^{1}H; ^{1}H,^{13}C-HSQC-DEPT (heteronuclear single-quantum correlation-distortionless enhancement by polarization transfer); ^{1}H, ^{1}H-COSY (correlation spectroscopy), ^{1}H, ^{1}H-TOCSY (total correlation spectroscopy)). Isomerization of *(all-E)*-rubixanthin was induced by exposition to sun-/daylight for 10 min. In addition to the aforementioned 2D experiments, selective 1D COSY, TOCSY and NOESY spectra were collected to assign the newly-formed signals in the aliphatic region. 3D lowest energy conformers were generated with Marv in Sketch (Version 17.1.2.0; ChemAxon Ltd., Budapest, Hungary) employing the MMFF94 force field [14].

2.3.5. Vitamin E Identification and Quantification

The chromatographic separation of the tocols was achieved using a normal-phase HPLC equipped with a fluorescence detector according to the method reported by Al-Yafeai et al. (2018) [3]. The identification of individual tocochromanols was achieved by comparing their retention times with those of external standards and tocols were quantified by 6-point calibration curve ($r^2 \geq 0.999$) of external standards. The total contents of vitamin E were estimated as the sum of all individual tocopherols and tocotrienols determined by HPLC analysis.

2.4. Vitamin C

2.4.1. Extraction Procedure

For vitamin C analysis, 0.5 ± 0.05 g of *R. rugosa* hips were extracted two times by using meta-phosphoric acid (4.5%). Following this, the samples were shaken for 1 min by vortex and centrifuged for 5 min at $3000 \times g$.

2.4.2. Quantification of Vitamin C

Vitamin C contents were determined based on reaction of 2,4-dinitrophenylhydrazine (DNP) with ketone group of dehydroascorbic acid under acidic conditions to form red osazone derivatives, which were subsequently measured at 540 nm. The experiment was performed in triplicate according to Al-Duais et al. (2009) [15]. The standard solutions of ascorbic acid were prepared in the range of 0.5–50 µg/mL. A standard curve of ascorbic acid reference was constructed and the ascorbic acid contents of *R. rugosa* hips were determined using the equation of the linear regression and are expressed as mg/100 g.

2.5. Antioxidants Capacity and Total Phenolic Contents

2.5.1. Extraction Procedures

Samples were prepared in triplicates, approximately 1 ± 0.05 g of *R. rugosa* hips were weighed into 15 mL centrifuge tubes. For hydrolysis, 1 mL hydrochloric acid (1 M) and glass beads were added to the hips. Afterwards, 1 mL sodium hydroxide (2 M in 75% MeOH) and 1 mL meta-phosphoric acid (0.75 M) were added. All steps were accompanied by shaking for 30 s and heating in a water bath (30 min 37 °C). Following the centrifugation (5 min, $3000 \times g$), the supernatant was collected in a 25 mL volumetric flask, while the residue was extracted again with 5 mL of 70% methanol, this step was accompanied by shaking for 1 min and centrifugation ($3000 \times g$ for 5 min). Finally, the extraction was repeated twice and for the final volume, the volumetric flask was filled up with 70% methanol. For the lipophilic test, analysis was performed with carotenoid extracts that have been mentioned above. The collected supernatants were concentrated in a rotary evaporator and the final volume was 5 mL. The resulting solutions were re-dried under a stream of nitrogen gas and re-dissolved in hexane. An aliquot of 1.5 mL was centrifuged again ($20000 \times g$, 5 min) and then different dilutions were prepared.

2.5.2. Determination of Total Phenolic Contents (TP)

Total phenolic contents, including other polar antioxidants (e.g., vitamin C), were assessed using the Folin-Ciocalteu method [16]. This method is based on a colorimetric oxidation/reduction reaction. Using a 96-well microtiter plate, 30 µL of HPLC water (blank), gallic acid monohydrate standard (8.51–170.12 mg/L) and methanol/water extract (70/30, *v/v*) were added. Following this, 150 µL of Folin-Ciocalteu's 1:10 diluted reagent and 120 µL of sodium carbonate solution (75 g/L) were gradually mixed with the samples. In darkness for 2 h and at ambient temperature, samples were incubated, the absorbance measurement was performed at 740 nm in the microplate reader at 30 °C. The TP is expressed as gallic acid equivalents (GAE) in mg/100 g.

2.5.3. Hydrophilic Trolox Equivalent Antioxidant Capacity (H-TEAC) Assay

The hydrophilic TEAC assays use intensely coloured cation radicals of ABTS$^{\bullet+}$ to test the ability of antioxidants to quench radicals and this was performed according to the method previously described [16]. The ABTS$^{\bullet+}$ radical was prepared using 10 mL ABTS solution (7 mM) and 10 mL of potassium peroxodisulfate solution (2.45 mM). In the darkness and at ambient temperature, the stock solution was incubated for 24 h, ABTS$^{\bullet+}$ working solution was freshly diluted with phosphate buffer (75 mM, pH 7.4). For analysis, in the 96-well microtiter plate, 20 µL of HPLC water (blank), Trolox standard (12.5–250 µM)) and methanol/water extract (70/30, *v/v*) were mixed with 200 mL ABTS$^{\bullet+}$ working solution. The drop of absorption of the stable radical ABTS$^{\bullet+}$ during the reaction with antioxidants is measured at 734 nm, the antioxidant capacity is expressed as Trolox equivalents (TE) in mmol/100 g.

2.5.4. Hydrophilic Oxygen Radical Absorbance Capacity (H-ORAC) Assay

In the H-ORAC assay, the reaction between the peroxyl radical and a fluorescent probe was achieved, generating a non-fluorescent product that can be readily quantified using fluorescence. For analysis, dilution 1:100 of the stock solution of fluorescein (0.12 mM) and phosphate buffer (75 mM, pH 7.4) was utilised to prepare a working solution of fluorescein (1.2 µM). The reaction was performed by adding 10 µL of methanol/water (70/30, *v/v*) extract, 25 µL working solution of fluorescein and 100 µL buffer into the 96-well plate wells. Following this, the microplate was incubated (10 min at 37 °C), after adding 150 µL of a freshly prepared AAPH solution (129 mM), the intensity of fluorescence was measured at 37 °C for 120 min every 60 s [16]. The antioxidant capacity was evaluated by reducing the rate of the non-fluorescent product formed over time. The protective effects of the antioxidants were calculated using the integrated area under the fluorescence decay curves (AUC). The Trolox equivalents (TE) are expressed as mmol TE/100 g.

2.5.5. α-Tocopherol Equivalent Antioxidants Capacity (TEAC) Assay

The α-TEAC method, previously described by Müller et al. 2010 [17], was used to evaluate the lipophilic antioxidant capacity in *R. rugosa* hips at different ripening times. Carotenoids extracts were prepared as described above and the final volume was 5 mL. Under a stream of nitrogen gas, the resulting solutions were dried and then re-dissolved in hexane. Meanwhile, ABTS$^{\bullet+}$ radical was produced by giving ABTS solution through a filter paper coated with manganese dioxide. The working solution of ABTS$^{\bullet+}$ was freshly prepared by diluting it with phosphate buffer (pH 7.4) and the absorbance measurement was carried out (0.70 ± 0.05) at 734 nm. The reaction was performed by adding 100 µL of lipophilic extract and 1000 µL ABTS$^{\bullet+}$ solution into the 2 mL cuvette, followed by mixing for 30 s. Phase separation was achieved through centrifugation (200× *g*, 30 s) and the absorbance was recorded exactly after 2 min. The α-tocopherol standard solutions were used to construct a linear regression line. The working standards of α-tocopherol (3.3–81 µmol/L) were prepared daily, linearity was given over the entire 5-point standard curve (r^2 = 0.999). The α-TEAC in the samples is expressed as mmol α-TE (tocopherol equivalents)/100 g.

2.6. Statistical Analysis

The data analysis was achieved using Prism program for windows, version 7.0 (Graph pad software, Inc, San Diego, CA, USA). All analyses were done in triplicate and the results are expressed as mean ± standard deviation (SD). A difference was considered statistically significant at $p < 0.05$. A one-way ANOVA (analysis of variance) followed by Student-Newman-Keuls post-hoc test (S-N-K) was utilised to compare data obtained from different analyses. Correlations were estimated using Pearson's correlation coefficient (R) and coefficient of determination (r^2) was used to determine the precision of methods used.

3. Results and Discussion

3.1. Carotenoids Identification and Quantification

Since most of the hips are harvested from wild plant groups, it is difficult to maintain quality aspects such as the contents of vitamins and antioxidants in raw materials during manufacturing processes. Although rosehips had more recently attracted attention because of their potential health benefits, there was little information about the changes of antioxidants properties that occur during the maturity especially in *R. rugosa*. Understanding physicochemical properties of fruits during the ripening is important to promote the levels of bioactive compounds through the selection of cultivar/species and harvest time. Due to the absence of esterified xanthophyll standards, the saponified extracts were utilised to identify and quantify carotenoids, although the damage of carotenoids cannot fully be avoided during saponification. The carotenoids characterisation and identification in the *R. rugosa* hips at different ripening degrees was achieved depending on their chemical properties and chromatographic and spectroscopic characteristics (UV-vis and MS). In addition, the mass spectrum and the abundance of molecular ion [M+H] + *m/z* results confirmed the previous findings [3].

Table 1. Concentrations of carotenoids (mg/100 g) in saponified extracts of *R. rugosa* at different ripening degrees.

Carotenoid \ Sample	Green	Orange	Red
(all-E)-β-carotene	1.8 ± 0.1 [a]	7.0 ± 0.4 [b]	7.3 ± 0.2 [b]
(9Z)-β-carotene	0.01± 0.0 [a]	n.d.	0.1 ± 0.01 [b]
(13Z)-β-carotene	n.d.	n.d.	0.3 ± 0.3
(15Z)-β-carotene	n.d.	n.d.	0.1 ± 0.01
(all-E)-α-cryptoxanthin	0.2 ± 0.01	0.2 ± 0.02	0.2 ± 0.01
(all-E)-β-cryptoxanthin	0.2 ± 0.02 [a]	0.4 ± 0.04 [b]	1.0 ± 0.03 [c]
(all-E)-lutein	0.5 ± 0.03	0.5 ± 0.0	0.5 ± 0.01
(all-E)-lycopene	1.6 ± 0.2 [a]	2.2 ± 0.4 [b]	6.0 ± 0.6 [c]
(5Z)-lycopene	0.4 ± 0.06 [a]	0.6 ± 0.08 [b]	1.4 ± 0.1 [c]
(9Z)-lycopene	1.2 ± 0.1 [b]	1.4 ± 0.2 [b]	0.5 ± 0.05 [a]
(13Z)-lycopene	4.2 ± 0.6 [a]	6.0 ± 0.9 [b]	6.5 ± 0.4 [b]
(all-E)-rubixanthin	0.9 ± 0.1 [a]	2.1 ± 0.2 [b]	3.0 ± 0.1 [c]
(5'Z)-rubixanthin	1.8 ± 0.04 [a]	n.d.	3.6 ± 0.3 [b]
(all-E)-violaxanthin	1.0 ± 0.3 [a]	1.2 ± 0.4 [b]	0.9 ± 0.4 [a]
(all-E)-zeaxanthin	0.5 ± 0.1 [a]	2.0 ± 0.1 [b]	1.7 ± 0.0 [b]
phytoene	0.8 ± 0.1 [c]	0.6 ± 0.03 [b]	0.4 ± 0.1 [a]
phytofluene	n.d.	n.d.	n.d.

Data are expressed as mean ± SD (*n* = 3). One-way ANOVA with Student-Newman-Keuls post-hoc test, different letters a/b/c in the same line indicate significant differences ($p < 0.05$), n.d. not detected.

The quantitative estimate (Table 1) showed variations between the concentrations of carotenoids in *R. rugosa* hips at different ripening degrees with significant difference ($p < 0.05$), the present study confirms our previous findings [3] in mature *R. rugosa* hips. Total carotenoids were analysed, during ripening the concentration of carotenoids increased due to the accumulations of *(all-E)*-β-carotene, *(all-E)*-rubixanthin and *(all-E)*-lycopene (Table 1). The findings of the current study are consistent with Fraser et al. (1994) [18] who suggested the concentration of carotenoids increased in tomato during ripening between 10- and 14-fold due mainly to the accumulation of lycopene (Figure 1). The total carotenoid contents were lowest in the immature stage and showed a pattern of continually increasing accumulation until the final stage of maturity, as was observed for lycopene. This is likely a result of the higher contribution of lycopene to total carotenoids, lycopene is a major carotenoid in mature rosehips [3,7]. On the other hand, Staffan et al. (2008) [19] reported that the total carotenoids content increased more than 10-fold during the eight weeks of maturation of *R. spinosissima*, while in other species of rosehips the carotenoids content increased 1.3–2.6 times during the five-week study period. Interestingly, in *R. rugosa* hips a significant increase in the total carotenoid contents (1.0-fold) was observed in orange colour, whereas a 1.5-fold increase was observed within mature stage. In the mature fruits, the carotenogenesis is controlled by particular mechanisms that differ from those in the photosynthetic tissues, therefore carotenoids are a fundamental part of the pigment-protein complexes in thylakoid [20]. This phenomenon has been attributed to an upregulation of carotenoid gene expression (phytoene synthase) with maturation. The enzyme catalyses the first committed step of carotenoid synthesis, the conversion of geranylgeranyl pyrophosphate to phytoene; phytoene serves as a precursor of lycopene from which several other carotenoid compounds are synthesized [21].

In accordance with these explanations, the present study has shown a significant decrease in the phytoene contents during the maturity (Table 1). Even though, xanthophylls such as violaxanthin, lutein, zeaxanthin and rubixanthin and carotenes such as β-carotene and lycopene were found at all stages of maturation, concentrations tended to change at different ripening stages. Likewise, (*all-E*)-β-carotene was intensively increased in the orange-red stages of maturity. In accordance with the present results, previous studies have demonstrated that higher (*all-E*)-β-carotene contents were also previously observed by Kotikova et al. (2011) [22] in the red stage in all commercial cultivars of tomatoes. (*all-E*)-β-Carotene played an important role as a photoprotective antioxidant during the photosynthesis procedure in immature fruits [23], as well as contributes to the colour of mature fruits together with lycopene [22]. Our study provides additional support for (*all-E*)- and (*Z*)-lycopene identification and quantification n rugosa hips at different ripening times. The HPLC chromatogram showed that (*all-E*)- and (*Z*)-lycopene isomers eluted at different retention times as shown in Figure 1, the peaks were determined as (*13Z*)-, (*9Z*)-, (*all-E*)- and (*5Z*)-lycopene.

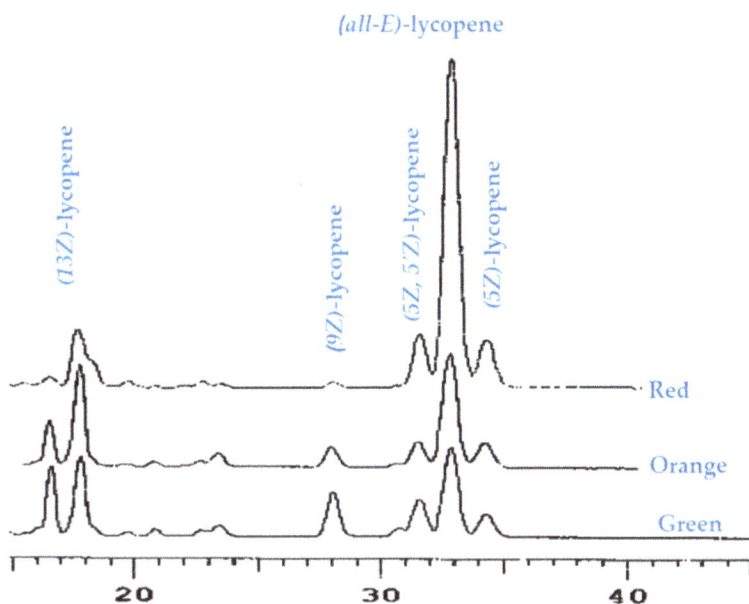

Figure 1. HPLC chromatograms of lycopene isomers in saponified extracts rosehips of *R. rugosa* at different ripening degrees, using a C30 column and an isocratic mobile phase.

These results support our findings [3] and are consistent with those of other studies indicating that *R. rugosa* hips contained three types of (*Z*)-lycopene [24]. The quantitative determination of (*Z*)-lycopene isomers is presented in Table 1, the statistical analysis revealed significant differences

($p < 0.05$). *(all-E)*-Lycopene contents showed an increasing pattern of accumulation based on total carotenoids during the ripening (Figure 2A).

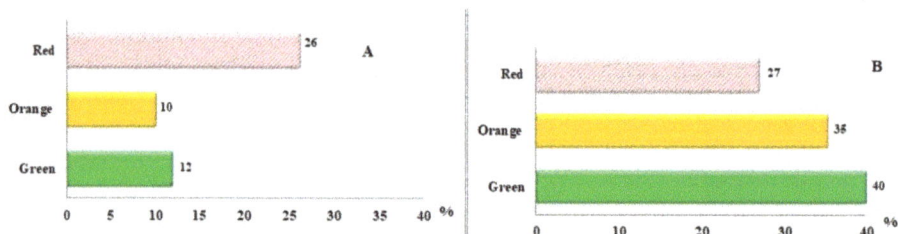

Figure 2. (**A**) The Percent (%) of *(all-E)*-lycopene based on total carotenoids in R. *rugosa* hips at different ripening degrees. (**B**) The Percent (%) of *(Z)*-lycopene based on total carotenoids in R. *rugosa* hips at different ripening degrees.

A possible explanation for this might be the transition of chloroplasts into chromoplasts. In contrast, *(Z)*-lycopene showed the opposite trend of accumulation compared to *(all-E)*-lycopene, such that its concentration decreased based on total carotenoids in fruits across the ripening stages (Figure 2B). In particular, the isolation of gazaniaxanthin from R. *rugosa* hips was problematic because of its low concentration. In addition, invisible compounds having a molecular weight of more than 1000 *m/z* were simultaneously eluted. For these reasons, the isomerization of *(all-E)* rubixanthin was used to confirm our results. The HPLC results obtained from the isomerization of *(all-E)*-rubixanthin by using $CDCl_3$ and sun light are set out in Figure 3A. Both pigments, *(all-E)*-rubixanthin and gazaniaxanthin, showed the spectra in HPLC mobile phase with maximum absorbance at 463.3 nm and 463.3 nm, with retention times of 33.17 and 34.35 min, respectively due to only slight differences in polarity. The UV spectrum was previously described and is characterized also by the appearance of a new maximum around 330–350 nm (cis-peak), ultraviolet spectrum is moved 5-10 nm towards the shorter wavelengths, and this finding supports our previous research [3] and is also comparable to recent results [25,26].

3.2. Isomerisation and NMR Analysis of (all-E)- and (5′Z)-Rubixanthin

(5′Z)-Rubixanthin (gazaniaxanthin) has been previously described as the main isomer for *(all-E)*-rubixanthin (Figure 3B) as well as being one of the important features that characterize R. *rugosa* compared to R. *canina* [3]. For further specific identification of (5′Z)-rubixanthin immediately after isomerization, the sample was checked by NMR technique. The ^{1}H spectra of *(all-E)*-rubixanthin before and after exposition to sunlight are presented in Figure 3C,D, basically two new signals appeared in the aliphatic region. Since the multiplet signal at 5.13 ppm (Figure 3C) belongs to the terminal CH proton of *(all-E)*-rubixanthin, we speculated that the new signal at 5.17 ppm (Figure 3D) was indicative of the rearrangement of the 5′ double bond from E to Z configuration. Based on selective COSY and TOCSY spectra that probe for neighbouring protons inside the same spin system, the second new signal of the light-induced isomer could be assigned to the CH_2 group terminal to the 5′ double bond (data not shown).

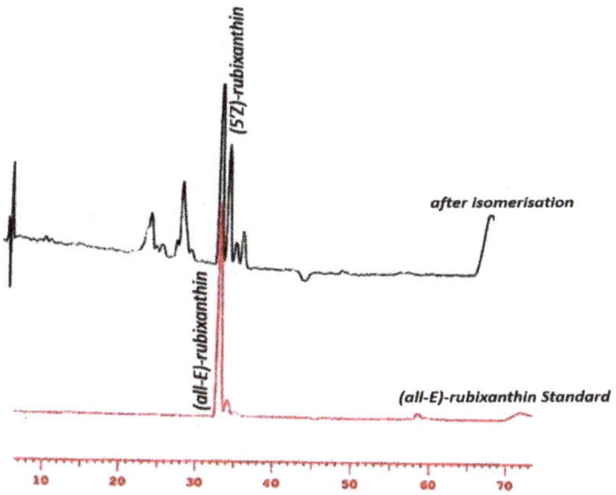

(A)

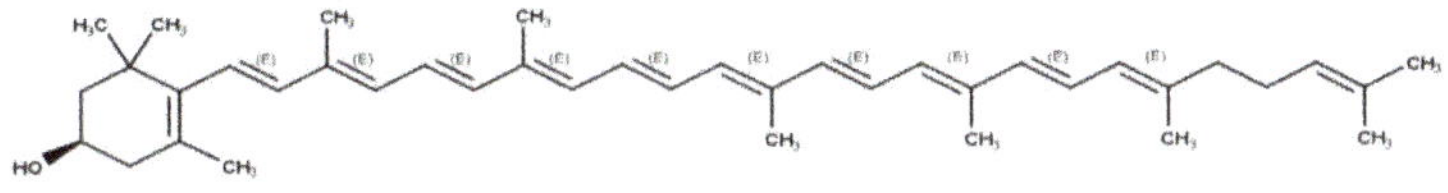

(*all-E*)-Rubixanthin

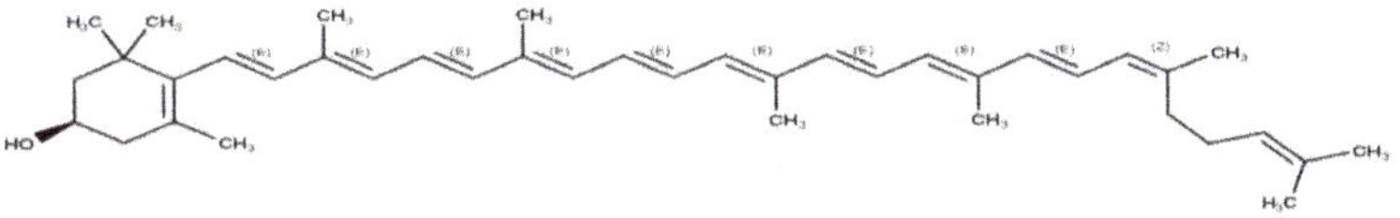

(5'Z)-Rubixanthin

(B)

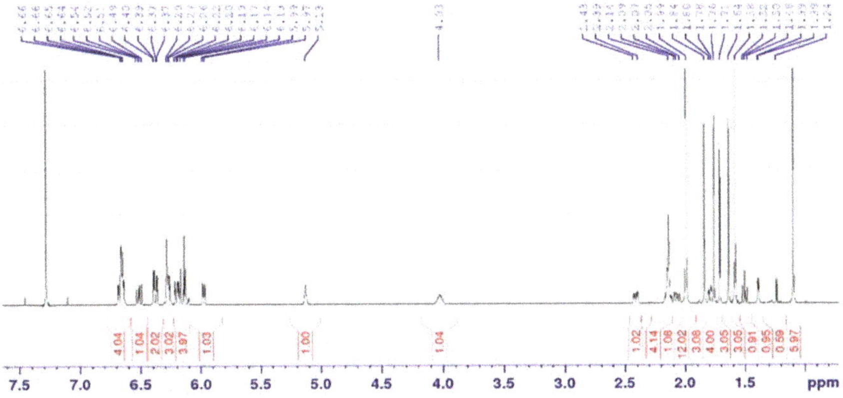

(C)

Figure 3. *Cont.*

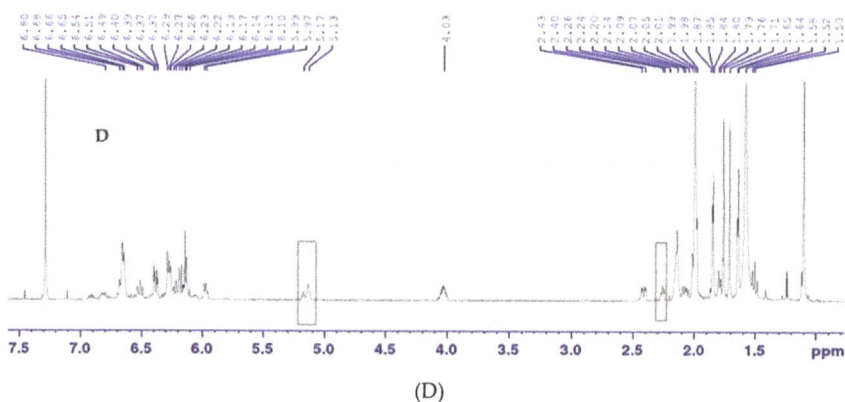

(D)

Figure 3. (**A**) Chromatographic separation of *(all-E)*-rubixanthin and *(5′Z)*-rubixanthin, using a C30 reversed phase column (see text for chromatographic conditions). (**B**) Structures of *(all-E)*-rubixanthin and *(5′Z)*-rubixanthin. (**C**) ^{1}H nuclear magnetic resonance (NMR) spectrum (600 MHz, 293 K, CDCl$_3$) of *(all-E)*-rubixanthin prior to light exposition. (**D**) ^{1}H NMR spectrum (600 MHz, 293 K, CDCl$_3$) of *(5′Z)*-rubixanthin after light exposition.

In the *(all-E)*-rubixanthin, both CH$_2$ groups between the 5′ double bond and the CH group have similar chemical/electronic environments resulting in a common signal at 2.14 ppm with an integral of 4 (Figure 3C). Upon light-induced isomerisation, the signal of the CH$_2$ group adjacent to the 5′ double bond slightly shifted downfield to 2.24 ppm. This shift can be attributed to the emerged spatial proximity of a CH$_3$ group upon isomerisation of rubixanthin (Figure 4). Taken together, based on the NMR data we reason that exposure of *(all-E)*-rubixanthin to sunlight results in an isomerisation of the very last double bond of the conjugated double bond system in the aliphatic chain of rubixanthin forming *(5′Z)*-rubixanthin (Figure 3B). The findings of the current study are consistent with those of [12,27] who suggested that gazaniaxanthin is the corresponding 5′-cis isomer of *(all-E)*-rubixanthin.

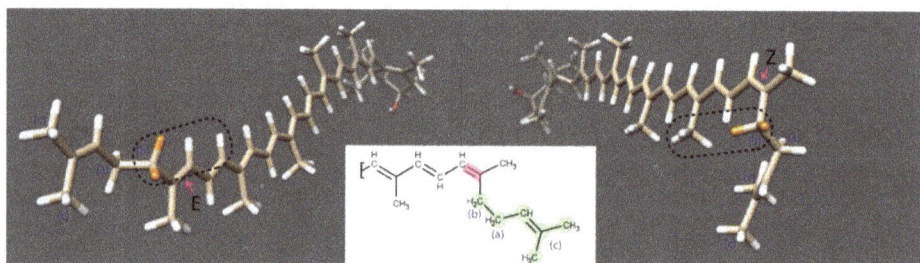

Figure 4. 3D structures of *(all-E)*-rubixanthin (left) and *(5′Z)*-rubixanthin (right), the 5′ double bond is highlighted in pink. The protons of CH$_2$ group adjacent to the 5′ double bond are marked in orange. As indicated by the dotted fields, in the *(all-E)*-rubixanthin this CH$_2$ group is exposed to CH protons only, whereas in the case of the *(5′Z)*-isomer it is additionally exposed to a methyl group.

3.3. Vitamin E Identification and Quantification

Vitamin E is an important lipid-soluble antioxidant found naturally in eight different forms, including α-, β-, γ-, δ-tocopherol and -tocotrienol. The basic structure of vitamin E is comprising a hydrophobic polyprenyl side chain with a polar chromanol ring, tocochromanols with a fully saturated side chain are called tocopherols and those with double bonds at positions 3′, 7′ and 11′ of the side chain are tocotrienols. The specific tocopherols and tocotrienols differ by number and positions of the methyl groups in the 6-chromanol ring, resulting in the α-, β-, γ- and δ-isomers (Figure 5) [28].

Tocopherol

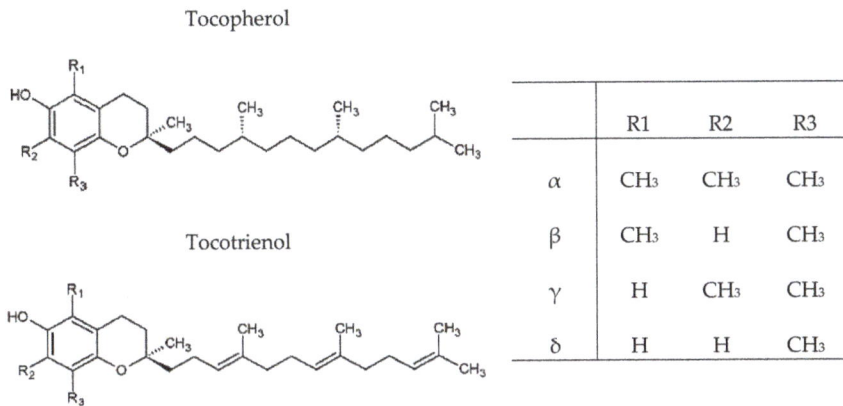

	R1	R2	R3
α	CH3	CH3	CH3
β	CH3	H	CH3
γ	H	CH3	CH3
δ	H	H	CH3

Tocotrienol

Figure 5. Chemical structures of tocopherols and tocotrienols.

α-Tocopherol is commonly referred to as vitamin E, because it is the most biologically active and widely distributed form in nature. Limited information is available concerning the contents of tocopherols and tocotrienols in rose hips especially in *R. rugosa*. To our knowledge, variations during the ripening times have not previously been investigated. In the present study, α-tocopherol was the main isomer of tocopherol in *R. rugosa* hips, this finding supports our previous research [3]. The content of α-tocopherol is likely to vary during the ripening period, the maximum concentration was achieved in the orange hips with significant difference ($p < 0.05$) (Table 2).

Table 2. Tocopherol and tocotrienol concentrations ($\mu mol/100$ g) in rosehips of *R. rugosa* at different ripening degrees.

Parameter \ Sample		Green	Orange	Red
α-Tocopherol		15 ± 3 [a]	17 ± 2 [b]	14 ± 1 [a]
γ-Tocopherol	$\mu mol/100$ g	n.d.	n.d.	n.d.
Σ Tocotrienol		n.d.	n.d.	n.d.
Σ Vitamin E		15 ± 3 [a]	17 ± 2 [b]	14 ± 1 [a]

Data are expressed as mean $\pm$ SD (n = 3). One-way ANOVA with Student-Newman-Keuls post-hoc test, different letters a/b within raw indicate significant differences ($p < 0.05$). n.d. = not detected.

The present findings seem to be consistent with Andersson et al. (2012) [29] who found the amounts of total tocopherols and vitamin E activity being decreased in rose hips during ripening, the change was relatively small and limited. Since α-tocopherol is the strongest tocopherol in quenching of singlet oxygen, the high concentrations of α-tocopherol participate in the stability of berries in the final stages of development [30].

3.4. Quantification of Vitamin C

Vitamin C (Ascorbic acid, AA) is the most powerful water-soluble antioxidant in human blood plasma, acts as a regenerator for vitamin E in lipid systems. AA is an odourless, white solid having the chemical formula $C_6H_8O_6$ [31]. An enhanced fruit AA pool has also been suggested to be associated with improved postharvest fruit quality in hard fruit species, such as pear and apple [32,33]. Generally, rosehips are considered the most abundant natural source of vitamin C, the contents ranged between 200 and 2800 mg/100 g [34]. Thus, the medicinal value of rosehips depends largely on the vitamin C contents. The results obtained from the preliminary analysis of AA in *R. rugosa* hips at different ripening degrees are shown in Table 3, the contents ranged between 798 mg/100 g and 1090 mg/100 g. It is encouraging to compare these findings with those found by Ercisli (2007) [35] who found that contents of AA in the fresh fruits of rose species were between 706 mg/100 g and 974 mg/100 g.

According to the results, the amounts of AA varied greatly at three maturity stages of *R. rugosa* hips with a significant difference ($p < 0.05$). The higher contents of vitamin C were achieved at the half-ripe (orange colour) stage with 14% of increase, whereas these contents decreased in the fully-ripe (red colour) stage by 16%. The present findings seem to be consistent with Zhang et al. (2006) [36] who found the AA contents increased at the beginning and middle of fruit growth and decreased after the development of fruits colour. Rousi & Aulin (1977) [37] reported a decreasing trend in vitamin C contents, accompanied by a steady increase in the fresh weight of the berries. On the other hand, the low vitamin C content in the plants may be due to the level of oxygen in the environment, the amount of light that reaches the plants and variations in endogenous plant growth regulators and temperature [38].

Table 3. Antioxidant capacity, total phenolic contents and ascorbic acid contents in *R. rugosa* hips at different ripening degrees.

Parameter \ Sample	Green	Orange	Red
Ascorbic acid (mg/100 g)	955 ± 71 [b]	1090 ± 51 [c]	798 ± 37 [a]
Total phenolics (mg GAE/100 g)	1097 ± 123 [a]	1009 ± 124 [a]	1327 ± 43 [b]
H-TEAC (mmol TE/100 g)	9.0 ± 1.0 [a]	9.1 ± 1.0 [a]	15 ± 1.0 [b]
H-ORAC (mmol TE/100 g)	32 ± 4 [b]	30 ± 4 [b]	23 ± 0.3 [a]
L-TEAC (mmol α-TE/100 g)	0.8 ± 0.2 [a]	1.1 ± 0.1 [a]	4.4 ± 0.3 [b]

Data are expressed as mean ± SD (n = 3). One-way ANOVA with Student-Newman-Keuls post-hoc test different letters a/b within row indicate significant differences between samples ($p < 0.05$). H: hydrophilic antioxidants, L: lipophilic antioxidants. GAE: gallic acid equivalents, TEAC: Trolox equivalent antioxidant capacity, ORAC: oxygen radical absorbance capacity, TE: Trolox equivalents, α-TE: α-tocopherol equivalents.

3.5. Antioxidant Capacity

The measurement of the antioxidant capacity of food products is a matter of growing interest because it may provide a variety of information, such as resistance to oxidation, quantitative contribution of antioxidant substances or the antioxidant capacity that they may present inside the organism when ingested [39]. Depending on the reaction mechanism the antioxidants capacity methods can be classified into two groups: hydrogen atom transfer (HAT) and electron transfer (ET). The majority of HAT-based assays measure the capability of an antioxidant to quench peroxyl radicals through transferred hydrogen atom (H) of a phenol (Ar-OH), which include the oxygen radical absorbance capacity (ORAC) assay. Whereas, ET-based assays measure the capacity of an antioxidant in the reduction of an oxidant, which changes colour when reduced. The degree of colour change is correlated with the sample's antioxidant concentrations, (ABTS)/hydrophilic Trolox-equivalent antioxidant capacity (H-TEAC) is one of the decolourisation assays [40].

3.5.1. Total Phenolic Contents (TP)

The antioxidant capacity of phenolic compounds is mainly due to their redox properties, which allow them to act as reducing agents, hydrogen donors, singlet oxygen quenchers or metal chelators [41]. Several previous studies have confirmed the presence of phenolic compounds in rosehips, Montazeri et al. (2011) [42] found that the levels of TP in *R. canina* extracts were 424.6 ± 1.8 mg GAE/g. In contrast, Denev et al. (2013) [43] reported that TP in the fruits of *R. canina* was 1934 ± 4 GAE/100 g. The maturity stage affects the TP in the *R. rugosa* hips, the changes in total phenolic contents during maturation are presented (Table 3). According to the results, the amounts of TP varied during maturity stages of *R. rugosa* hips with a statistically significant increase ($p < 0.001$) between maturity stages. Furthermore, the higher contents of TP were achieved at the ripe stage (red colour) with 21% of increase. TP accumulation in plants can be affected by genetic factors, environmental and cultural conditions as well as by various stresses [44]. These factors can explain the data differences reported within the scientific studies.

3.5.2. Hydrophilic Antioxidant Capacity

During maturity, many biochemical, physiological and structural modifications occur, influencing the content of phytochemicals and thus affecting the antioxidant capacity. To the best of our knowledge, there are no previous studies to determine the variation of total antioxidant capacity in *R. rugosa* hips during ripening. Assays to measure total antioxidant capacity can be direct, which are based on the ability to inhibit the oxidation of a substance. The most common direct assays are H-TEAC and H-ORAC. As shown in Table 3, an increasing trend in the H-TEAC of *R. rugosa* hips was observed with significant differences ($p < 0.001$) between the different maturity stages. The highest activities were achieved at full maturity (red colour), the rate of increase was 67%. A possible explanation for these results is a correlation between bioactive compounds and antioxidant activities and this seems to be consistent with research done by [45]. Further analysis showed a positive correlation between TP and H-TEAC in *R. rugosa* hips during maturity (r = 0.9594). The observed increase in H-TEAC could be attributed to the contribution of predominant types of phenolic compounds at different ripening stages to overall antioxidant capacity. In the same vein, Müller et al. (2010) [16] suggested the major portion of antioxidant capacity was generated by polyphenolic compounds. On the other hand, the antioxidant capacity measured by the H-ORAC assay showed a significant variation ($p < 0.001$) as well. In fact, a decrease of H-ORAC values has been observed as the ripeness increased. These results may be explained by the fact that Folin-Ciocalteu and the radical scavenging method ABTS share the same reaction mechanism (electron transfer), whereas the H-ORAC method is based on hydrogen atom transfer reactions. The present findings seem to be consistent with Rodríguez et al. (2016) [46] who found the antioxidants capacity measured by H-ORAC significantly decreased as the palm fruit became ripe.

3.5.3. Lipophilic Antioxidant Capacity

As mentioned above, rosehips are rich in lipid-soluble antioxidants such as carotenoids and tocopherols. The most abundant carotenoids were in the following order *(all-E)-β*-carotene, *(all-E)-* and *(Z)*-lycopene, followed by *(all-E)-* and *(Z)*-rubixanthin, *(all-E)*-zeaxanthin and *(all-E)-β*-cryptoxanthin. Among the different defence strategies, carotenoids are efficient deactivators of electronically excited sensitizer molecules that contribute to the generation of radicals and singlet oxygen. Moreover, carotenoids are involved in the quenching of reactive oxygen species, the molecular oxygen ($^1O_2$). For the physical scavenging, the efficacy of carotenoids is not only associated with the number of conjugated double bonds present in the molecule but also with the type of ring, functional groups on the rings, and so forth. [47]. Since vitamin E is a lipophilic molecule, its antioxidant functions are important for the protection of membrane lipids against peroxidation. α-Tocopherol is a chain breaking antioxidant capable of interrupting radical chain reactions produced by the lipid peroxyl radicals (LOO•). In addition, it directly scavenges superoxide radicals and singlet oxygen. As an antioxidant, the tocopherol reacts very quickly with peroxyl radicals to tocopheroxyl radicals due to its 6-hydroxychroman structure before it can abstract hydrogen from a target. The antioxidant capacity of the lipophilic extracts was determined using the L-TEAC method [17]. As a result, an increase in antioxidant capacity of lipophilic extracts (Table 3) was observed with statistical significance ($p < 0.0001$). The increase was progressed during maturity with +38% in orange colour and with +450% at full maturity (red colour). This finding confirms our previous results of accumulation of carotenoids during ripening.

4. Conclusions

This research extends our knowledge of rosehips, confirms previous findings and contributes additional evidence that suggests rosehips are a good source of carotenoids, vitamin E and vitamin C. Maturity stage affects the bioactive compounds as well as the antioxidant capacity in the fruit. Late harvesting is recommended if a high content of carotenoids is desire, while harvesting should be carried out earlier if higher vitamin E and vitamin C contents are desired.

Author Contributions: Conceptualization, A.A.-Y. and V.B.; Investigation, A.A.-Y. and P.B.; Writing-Original Draft Preparation, A.A-Y.; Writing-Review & Editing, A.A.-Y., P.B. and V.B.

Funding: This research did not receive any specific grant from funding agencies in the public, commercial or not-for-profit sectors

Acknowledgments: We greatly appreciate DSM (Kaiseraugst, Switzerland) for providing the *(all-E)*-rubixanthin standard.

Conflicts of Interest: The authors declare no existing conflict of interest.

References

1. Singh, P.; Goyal, G.K. Dietary lycopene: Its properties and anticarcinogenic effects. *Compr. Rev. Food Sci. Food Saf.* **2008**, *7*, 255–270. [CrossRef]
2. Nilsson, O. Rosa. In *Flora of Turkey and the East Aegean Islands*, 4th ed.; Davis, P.H., Ed.; Edinburgh University Press: Edinburgh, UK, 1997; pp. 106–128.
3. Al-Yafeai, A.; Malarski, A.; Böhm, V. Characterization of carotenoids and vitamin E in *R. rugosa* and *R. canina*: Comparative analysis. *Food Chem.* **2018**, *242*, 435–442. [CrossRef] [PubMed]
4. Andersson, S.C.; Rumpunen, K.; Johansson, E.; Olsson, M.E. Carotenoid content and composition in rose hips (*Rosa* spp.) during ripening, determination of suitable maturity marker and implications for health promoting food products. *Food Chem.* **2011**, *128*, 689–696. [CrossRef]
5. Chrubasik, C.; Roufogalis, B.D.; Müller-Ladner, U.; Chrubasik, S. A systematic review on the *Rosa canina* effect and efficacy profiles. *Phytother. Res.* **2008**, *22*, 725–733. [CrossRef] [PubMed]
6. Gao, X.; Bjork, L.; Trajkovski, V.; Uggla, M. Evaluation of antioxidant activities of rosehip ethanol extracts in different test systems. *J. Sci. Food Agric.* **2000**, *80*, 2021–2027. [CrossRef]
7. Böhm, V.; Fröhlich, K.; Bitsch, R. Rosehip-a "new" source of lycopene. *Mol. Aspects. Med.* **2003**, *24*, 385–389. [CrossRef]
8. Daels-Rakotoarison, D.A.; Gressier, B.; Trotin, F.; Brunet, C.; Luyckx, M.; Dine, T.; Cazin, J.-C. Effects of *Rosa canina* fruit extract on neutrophil respiratory burst. *Phytother. Res.* **2002**, *16*, 157–161. [CrossRef] [PubMed]
9. Guo, D.; Xu, L.; Cao, X.; Guo, Y.; Chan, C.-O.; Mok, D.K.W.; Yu, Z.; Chen, S. Anti-inflammatory activities and mechanisms of action of the petroleum ether fraction of *Rosa multiflora* Thunb. hips. *J. Ethnopharmacol.* **2011**, *138*, 717–722. [CrossRef] [PubMed]
10. Barros, L.; Carvalho, A.M.; Ferreira, I.C.R. Exotic fruits as a source of important phytochemicals: Improving the traditional use of *Rosa canina* fruits in Portugal. *Food Res. Int.* **2011**, *44*, 2233–2236. [CrossRef]
11. Müller, L.; Goupy, P.; Fröhlich, K.; Dangles, O.; Caris-Veyrat, C.; Böhm, V. Comparative study on antioxidant activity of lycopene (Z)-isomers in different assays. *J. Agric. Food Chem.* **2011**, *59*, 4504–4511. [CrossRef] [PubMed]
12. Arpin, N.; Liaaen-Jensen, S. Carotenoids of higher plants-11 Rubixanthin and Gazaniaxanthin. *Phytochemistry* **1969**, *8*, 185–193. [CrossRef]
13. Kishimoto, S.; Maoka, T.; Sumitomo, K.; Ohmiya, A. Analysis of carotenoid composition in petals of calendula (*Calendula officinalis* L.). *Biosci. Biotechnol. Biochem.* **2005**, *69*, 2122–2128. [CrossRef] [PubMed]
14. Halgren, T.A. Merck molecular force field. I. Basis, form, scope, parameterization, and performance of MMFF94. *J. Comput. Chem.* **1996**, *17*, 490–519. [CrossRef]
15. Al-Duais, M.; Hohbein, J.; Werner, S.; Böhm, V.; Jetschke, G. Contents of vitamin C, carotenoids, tocopherols, and tocotrienols in the subtropical plant species Cyphostemma digitatum as affected by processing. *J. Agric. Food Chem.* **2009**, *57*, 5420–5427. [CrossRef] [PubMed]
16. Müller, L.; Gnoyke, S.; Popken, A.; Böhm, V. Antioxidant capacity and related parameters of different fruit formulations. *LWT—Food Sci. Technol.* **2010**, *43*, 992–999. [CrossRef]
17. Müller, L.; Theile, K.; Böhm, V. In vitro antioxidant activity of tocopherols and tocotrienols and comparison of vitamin E concentration and lipophilic antioxidant capacity in human plasma. *Mol. Nutr. Food Rer.* **2010**, *54*, 731–742. [CrossRef] [PubMed]

18. Fraser, P.D.; Truesdale, M.R.; Bird, C.R.; Schuch, W.; Bramley, P.M. Carotenoid biosynthesis during tomato fruit development. *Plant Physiol.* **1994**, *105*, 405–413. [CrossRef] [PubMed]

19. Staffan, C.A.; Kimmo, R.; Eva, J.; Marie, E.O. Changes in tocopherols and carotenoids during ripening in rose hips and sea buckthorn berries. *J. Plan. Food Medic.* **2008**, *765*, 255–262. [CrossRef]

20. Thelander, M.; Narita, J.O.; Gruissem, W. Plastid differentiation and pigment biosynthesis during tomato fruit ripening. *J. Plant Biochem. Physiol.* **1986**, *5*, 128–141.

21. Li, L.; Yuna, H. Chromoplast biogenesis and carotenoid accumulation. *J. Arch. Biochem. Biophys.* **2013**, *539*, 102–109. [CrossRef] [PubMed]

22. Kotíkova, Z.; Lachman, J.; Hejtmankova, A.; Hejtmankov, K. Determination of antioxidant activity and antioxidant content in tomato varieties and evaluation of mutual interactions between antioxidants. *LWT—Food Sci. Technol.* **2011**, *44*, 1703–1710. [CrossRef]

23. Cazzaniga, S.; Bressan, M.; Carbonera, D.; Agostini, A.; Dall'Osto, L. Differential roles of carotenes and xanthophylls in photosystem I photoprotection. *J. Biochem.* **2016**, *55*, 3636–3649. [CrossRef] [PubMed]

24. Zhong, L.; Gustavsson, K.E.; Oredsson, S.; Gła, B.; Yilmaz, J.L.; Olsson, M.E. Determination of free and esterified carotenoid composition in rosehip fruit by HPLC DAD-APCI+-MS. *J. Food Chem.* **2016**, *210*, 541–550. [CrossRef] [PubMed]

25. Honda, M.; Takahashi, N.; Kuwa, T.; Takehara, M.; Inoue, Y.; Kumagai, T. Spectral characterisation of Z-isomers of lycopene formed during heat treatment and solvent effects on the E/Z isomerisation. *J. Food Chem.* **2015**, *171*, 323–329. [CrossRef] [PubMed]

26. Huawei, Z.; Xiaowen, W.; Elshareif, O.; Hong, L.; Qingrui, S.; Lianfu, Z. Isomerisation and degradation of lycopene during heat processing in simulated food system. *Food Res. Int. J.* **2014**, *21*, 45–50.

27. Schön, K. Studies on Carotenoids V. Gazaniaxanthin. *Biochem. J.* **1938**, *32*, 1566–1570. [CrossRef] [PubMed]

28. Muné-Bosch, S.; Alegre, L. The function of tocopherols and tocotrienols in plants. *CRC Crit. Rev. Plant Sci.* **2002**, *21*, 31–57. [CrossRef]

29. Andersson, S.; Olsson, M.; Gustavsson, K.; Johanssonc, E.; Rumpunenb, P. Tocopherols in rose hips (Rosa spp.) during ripening. *J. Sci. Food Agric.* **2012**, *92*, 2116–2121. [CrossRef] [PubMed]

30. Zadernowski, R.; Naczk, M.; Amarowicz, R. Tocopherols in Sea Buckthorn (*Hippophae rhamnoides* L.) Berry Oil. *JAOCS* **2003**, *80*, 55–58. [CrossRef]

31. Groff, J.L.; Gropper, S.S.; Hunt, S.M. The water soluble vitamins. In *Advanced Nutrition and Human Metabolism*; West Publishing Company: Minneapolis, MN, USA, 1995; pp. 222–231.

32. Franck, C.; Baetens, M.; Lammertyn, J.; Verboven, P.; Davey, M.W.; Nicolai, B.M. Ascorbic acid concentration in cv. conference pears during fruit development and postharvest storage. *J. Agric. Food Chem.* **2003**, *51*, 4757–4763. [CrossRef] [PubMed]

33. Davey, M.W.; Auwerkerken, A.; Keulemans, J. Relationship of apple vitamin C and antioxidant contents to harvest date and postharvest pathogen infection. *J. Sci. Food Agric.* **2007**, *87*, 802–813. [CrossRef]

34. Uggla, M.; Gao, X.; Werlemark, G. Variation among and within dog rose taxa (Rosa sect. Caninae) in fruit weight, percentages of fruit flesh and dry matter, and vitamin C content. *Acta Agric. Scand. B-Soil Plant Sci.* **2003**, *53*, 147–155.

35. Ercisli, S. Chemical composition of fruits in some rose (Rosa spp.) species. *J. Food Chem.* **2007**, *104*, 1379–1384. [CrossRef]

36. Zhang, J.G.; Luo, H.M.; Huang, Q.J.Y.; Shan, W.C.Y. A comparative study on berry characteristics of large berry cultivars of sea buckthorn. *For. Res. Beijing* **2006**, *18*, 643–650.

37. Rousi, A.; Hannele, A. Ascorbic acid contents in relation to ripeness in fruit of six Hippophae rhamnoides clones form pyharanta, SW. Finland. *J. Ann. Agric. Fenn.* **1977**, *16*, 80–87.

38. Seung, K.L.; Adel, A. Kade Preharvest and postharvest factors influencing vitamin C content of horticultural. *Postharvest Biol. Technol.* **2000**, *20*, 207–222.

39. Prior, R.L.; Hoang, H.; Gu, L.; Wu, X.; Bacchiocca, M.; Howard, L.; Hampsch-Woodill, M.; Huang, D.; Ou, B.; Jacob, R. Assays for Hydrophilic and Lipophilic Antioxidant Capacity (oxygen radical absorbance capacity (ORACFL)) of Plasma and Other Biological and Food Samples. *J. Agric. Food Chem.* **2003**, *51*, 3273–3279. [CrossRef] [PubMed]

40. Huang, D.; Ou, B.; Prior, R.L. The Chemistry behind Antioxidant Capacity Assays. *J. Agric. Food Chem.* **2005**, *53*, 1841–1856. [CrossRef] [PubMed]

41. Balasundram, N.; Sundram, K.; Sammar, S. Phenolic compounds in plants and agro-industrial by-products: Antioxidant Activity. Occurrence and Potential Uses. *J. Food Chem.* **2006**, *99*, 191–203. [CrossRef]

42. Montazeri, N.; Baher, E.; Mirzajani, F.; Barami, Z.; Yousefian, S. Phytochemical contents and biological activities of *Rosa canina* fruit from Iran. *J. Med. Plant Res.* **2011**, *5*, 4584–4589.

43. Denev, P.; Lojek, A.; Ciz, M.; Kratchanova, M. Antioxidant activity and polyphenol content of Bulgarian fruits. *J. Agric. Sci.* **2013**, *19*, 22–27.

44. Parr, A.J.; Bolwell, G.P. Phenols in the plant and in man. The potential for possible nutritional enhancement of the diet by modifying the phenols content or profile. *J. Sci. Food Agric.* **2009**, *80*, 985–1012. [CrossRef]

45. Ilahy, R.; Hdider, C.; Lenucci, M.S.; Tlili, I.; Dalessandro, G. Antioxidant activity and bioactive compound changes during fruit ripening of high-lycopene tomato cultivars. *J. Food Compost. Anal.* **2001**, *24*, 588–595. [CrossRef]

46. Rodríguez, J.C.; Gómez, D.; Pacetti, D.; Núñez, O.; Gagliardi, R.; Frega, N.; Ojeda, M.; Loizzo, M.; Tundis, R.; Lucci, P. Effects of the fruit ripening stage on antioxidant capacity, total phenolics, and polyphenolic composition of crude palm oil from interspecific hybrid Elaeis oleifera × Elaeis guineensis. *J. Agric. Food Chem.* **2016**, *64*, 852–859. [CrossRef] [PubMed]

47. Young, A.J.; Lowe, G.M. Antioxidant and prooxidant properties of carotenoids. *Arch. Biochem. Biophys.* **2001**, *385*, 20–27. [CrossRef] [PubMed]

antioxidants

MDPI

Article

In Vitro Antioxidant, Antityrosinase, and Cytotoxic Activities of Astaxanthin from Shrimp Waste

Sutasinee Chintong [1], Wipaporn Phatvej [2], Ubon Rerk-Am [2], Yaowapha Waiprib [1] and Wanwimol Klaypradit [1,3,*]

[1]   Department of Fishery Products, Faculty of Fisheries, Kasetsart University, Bangkok 10900, Thailand; sutasinee_c@yahoo.com (S.C.); ffisywp@gmail.com (Y.W.)

[2]   Expert Center of Innovative Herbal Products, Thailand Institute of Scientific and Technological Research, Pathum Thani 12120, Thailand; wipaporn@tistr.or.th (W.P.); ubon@tistr.or.th (U.R.-A.)

[3]   Center for Advanced Studies for Agriculture and Food, Kasetsart University Institute for Advanced Studies, Kasetsart University, Bangkok 10900, Thailand

*   Correspondence: ffiswak@ku.ac.th; Tel.: +66-2-9428644-5

Received: 6 March 2019; Accepted: 9 May 2019; Published: 13 May 2019

Abstract: Astaxanthin is a potent antioxidant compared with vitamins and other antioxidants. However, astaxanthin extract from shrimp processing waste has not yet been used in cosmetic products. This study aimed to explore the natural astaxanthin from shrimp shells for antioxidant and antityrosinase activities as well as potential toxicity. The antioxidant activities were performed with 2,2-diphenyl-1-picrylhydrazyl (DPPH), 2,2′-azino-bis(3-ethylbenzothiazoline-6-sulfonic acid) (ABTS) radical scavenging, β-carotene bleaching, and singlet oxygen quenching assays. The results revealed that astaxanthin extract demonstrated potent antioxidant activities against DPPH and ABTS radicals, and prevented the bleaching of β-carotene and quenching of singlet oxygen (EC_{50} 17.5 ± 3.6, 7.7 ± 0.6, 15.1 ± 1.9 and 9.2 ± 0.5 μg/mL, respectively). Furthermore, the astaxanthin extract could inhibit tyrosinase activity (IC_{50} 12.2 ± 1.5 μg/mL) and had no toxic effects on human dermal fibroblast cells. These results suggested that shrimp astaxanthin would be a promising dietary supplement for skin health applications.

Keywords: shrimp astaxanthin; antioxidant property; tyrosinase inhibition; cytotoxicity

---

1. Introduction

Astaxanthin is a ketocarotenoid synthesized by plants and microorganisms but is distributed mainly in aquatic animals such as crustaceans, salmon, and trout. It possesses antioxidant activity 10 times stronger than zeaxanthin, lutein, and canthaxanthin [1] and has beneficial effects supporting human health including alleviation of oxidative stress, inhibition of low-density lipoprotein (LDL) oxidation, enhancement of immune response, anti-inflammatory and anti-aging properties, with its major mode of action being a scavenger of reactive oxygen species (ROS) [2].

Shrimp waste is one of the important natural sources of carotenoid; astaxanthin and its esters as the major pigments. Several methods for extraction of astaxanthin have been attempted. Organic solvent has been used to recover the pigment from crustacean processing discards [3]. Alcohol was considered to be appropriate for astaxanthin due to its safety, effectiveness, and easier separation. Ethanolic extract of astaxanthin was used to fortify in yogurt as an optional functional food for consumers [4] and was investigated the anti-inflammatory effect on alveolar macrophages [5]. Nevertheless, the extract from shrimp waste has not been yet studied on skin effect.

Some studies have been done on skin protective effects by astaxanthin from *Haematococcus pluvialis*. These studies have demonstrated that astaxanthin contributed to maintain moisture of skin and decrease skin roughness, an early stage of wrinkle formation, by inhibiting lipid peroxidation [6] and also effectively inhibiting UVB-induced age spots and atopic dermatitis [7]. Moreover, the effects of astaxanthin in ROS scavengers or inhibitors have been implicated in the treatment of skin pigmentation disorders [8]. Skin pigmentation, such as suntan, blemishes, and freckles, is caused by excessive and uneven production of melanin. It is known that pigmentation in the skin can be prevented by reducing tyrosinase activity, either by inhibiting the synthesis of tyrosinase—which is an important enzyme for melanin synthesis—or by using an antagonist of the substrate for tyrosinase. Pigmentation is also prevented by inhibiting auto-oxidation of dopa and suppressing inflammatory reactions [9].

Previous reports showed that astaxanthin could reduce tyrosinase activity and prevent melanin synthesis by inhibiting auto-oxidation of dopa and dopaquinone, and as a result, the amount of melanin was decreased by 40% with no change in cell viability of B16 mouse melanoma cells [10]. Also, it was found that astaxanthin attenuates the stem cell factor (SCF), which downregulates tyrosinase activity and reduces melanin production [11]. In vivo study demonstrated multiple biological activities of astaxanthin to preserve skin health and achieve effective skin cancer chemoprevention [12]. In addition, with respect to safety, no adverse events were observed with oral astaxanthin supplementation [13].

These studies suggest skin therapeutic potentials of astaxanthin; however, there is currently no report showing the utilization of astaxanthin from shrimp waste, which is a natural and cheap source for astaxanthin recovery [14]. At present, natural astaxanthin derived from *Haematococcus pluvialis* is the major dietary supplements for humans and animals [15].

In the course of screening such natural substances, in vitro studies must be conducted to determine their efficacy and safety. Therefore, the aim of this study was to investigate the concentration of astaxanthin from shrimp shells which exhibits optimal levels of antioxidant and antityrosinase activities as well as to evaluate the possible cytotoxic effects on human dermal fibroblast cells in order to safely utilize as a functional ingredient for skin health.

2. Materials and Methods

2.1. Materials

Fresh white shrimp shells (*Litopenaeus vannamei*) were obtained from a frozen shrimp processing plant in Samut-Sakorn Province, Thailand. The materials were packed into plastic food storage bags with the weight approximately 1 kg and frozen at −20 °C until use. Prior to use, samples were thawed by submerging the bag under running tap water until completely thawed.

2.2. Preparation of Astaxanthin from Shrimp Shells

Astaxanthin in shrimp waste was extracted using ethanol. Five hundred grams of shrimp shells was blended with 1000 mL of ethanol using a Waring laboratory blender (Waring Laboratory Science, Winsted, USA). Shrimp shell residues were vacuum-filtered and evaporated under vacuum at 40 °C, 175 MPa using a rotary evaporator (Büchi Labortechnik AG, Flawil, Switzerland). The resulting concentrate was stored at −20 °C until further analysis.

2.3. Analysis of Astaxanthin Extract

The content of astaxanthin from shrimp shells was performed using HPLC analysis as previously described [16]. The astaxanthin extract was dissolved in a dichloromethane/methanol (HPLC grade) (25:75) mixture. The sample was passed through a 0.45 μm filter. Twenty microliters of filtered sample was injected into the HPLC system (LC 20A, Shimadzu, Japan) and eluted through a reversed phase C18 column (Beckman Ultrasphere C18, 250 × 4.6 mm, Phenomenex, Torrance, CA, USA). Chromatography was isocratically performed at 25 °C. The mobile phase consisted of methanol, dichloromethane, acetonitrile, and water (85:5:5:5 *v/v/v/v*), at a flow rate of 1 mL/min. The UV detection of elute was performed at 480 nm. Astaxanthin was qualitatively analyzed by comparing the retention times of standards and their quantifications were done by using a calibration curve.

2.4. Antioxidant Properties Assay

2.4.1. Determination of DPPH Free Radical-Scavenging Activity

The radical scavenging activity of astaxanthin was evaluated using a modified 2, 2-diphenyl-1-picryl-hydrazyl (DPPH) radical scavenging assay following previous report [17]. One hundred microliters of the DPPH solution in methanol was added to 100 µL of various concentrations of the astaxanthin in methanol (3–50 µg/mL), then shaken vigorously and allowed to stand in the dark at room temperature for 30 min. The absorbance of the sample solution was measured at 517 nm using a microplate reader (Thermo Fisher Scientific, Inc., Waltham, MA, USA). A curve of astaxanthin concentration against %DPPH was generated to estimate the concentration of astaxanthin needed to cause a 50% reduction of the initial DPPH concentration. This value is known as EC_{50} (efficient concentration when 50% oxidation is achieved, also called oxidation index) and was expressed in units of µg/mL. This assay was carried out in triplicate and the mean values were used to calculate the EC_{50}. Ascorbic acid and BHT (butylated hydroxytoluene) were used as positive controls. The scavenging effect on the DPPH inhibition as percentage (%) was calculated according to the equation

$$\%DPPH\ radical\ scavenging = [A_{control} - (A_{sample} - A_{sample\ blank})/A_{control}] \times 100, \qquad (1)$$

where $A_{control}$, A_{sample}, and $A_{sample\ blank}$ are the absorbance of the DPPH solution in methanol, astaxanthin solution with DPPH, and astaxanthin solution without DPPH, respectively.

2.4.2. Determination of ABTS Radical Scavenging Activity

The ABTS radical scavenging assay was measured following the modified method as previously described [18]. The ABTS radical solution was prepared by mixing 7 mM ABTS and 2 mM $K_2S_2O_8$ in equal quantities and incubating at room temperature for 16 h in the dark. The solution was diluted with water to obtain an absorbance of 0.70 ± 0.03 units at 734 nm. Two hundred microliters of the working ABTS solution was mixed with 20 µL of the different concentrations of the extract (astaxanthin in ethanol 3–50 µg/mL), the mixture was incubated at room temperature for 6 min, and then absorbance was recorded at 734 nm. Ascorbic acid and BHT were used as positive controls. All determinations were carried out in triplicate and EC_{50} value was expressed as the sample concentration which can quench fifty percent of ABTS radicals. The percentage of ABTS radical scavenging was calculated as

$$\%ABTS\ radical\ scavenging = [A_{control} - (A_{sample} - A_{sample\ blank})/A_{control}] \times 100, \qquad (2)$$

where $A_{control}$, A_{sample}, and $A_{sample\ blank}$ are the absorbance of the ABTS solution, astaxanthin solution with ABTS, and astaxanthin solution without ABTS, respectively.

2.4.3. Determination of β-Carotene Bleaching (BCB) Activity

Protocol for BCB activity was previously described with slight modifications [19]. A solution of β-carotene/linoleic acid was initially prepared by dissolving 10 mg of β-carotene in 10 mL of chloroform. Two mL of this solution was added to 40 mg of linoleic acid and 400 mg of Tween 40. The chloroform was evaporated off using a rotary evaporator. Then, 100 mL of aerated distilled water was added to the mixture with vigorous shaking. Aliquots of 100 µL of β-carotene/linoleic acid emulsion were added to 100 µL of different concentrations of astaxanthin in ethanol (3–50 µg/mL) in a 96-well plate. The zero time absorbance was measured at 470 nm using a microplate reader. The plates were incubated at 50 °C in a water bath and measurement of absorbance was recorded after 60 min; a blank, absent of β-carotene, was prepared for background deduction. The same procedure was repeated with the synthetic antioxidants, ascorbic acid, and BHT as positive controls. The assay was carried out in triplicate and EC_{50} value was calculated. The percentage antioxidant activity (%AOA) was calculated using the formula

$$\%AOA = [(Ac_{(0)} - Ac_{(60)})/(As_{(0)} - As_{(60)})]/(Ac_{(0)} - Ac_{(60)}) \times 100, \quad (3)$$

where $Ac_{(0)}$ is the absorbance of the control at $t = 0$ min, $Ac_{(60)}$ is the absorbance of the control at $t = 60$ min, $As_{(0)}$ is the absorbance of the sample at $t = 0$ min, and $As_{(60)}$ is the absorbance of the sample at $t = 60$ min.

2.4.4. Determination of Singlet Oxygen Scavenging ($^1O_2$) Activity

The formation of singlet oxygen was determined by monitoring p-nitrosodimethylaniline (RNO) bleaching according to a modified method as was previously described [20]. The bleaching of RNO was measured after singlet oxygen was generated by reacting H_2O_2 and NaOCl. The reaction mixture contained 50 μL of sample at astxanthin in ethanol various concentrations (3–50 μg/mL), 50 μL of 45 mM phosphate buffer (pH 7.1), 50 μL of 50 mM H_2O_2, 50 μL of NaOCl, 50 mM, 50 μL of 50 mM histidine, and 50 μL of 10 μM RNO. The mixture was incubated at 30 °C for 40 min, and the decrease in RNO absorbance was measured at 440 nm. Rutin and quercetin were used as positive controls and the values were expressed as EC_{50}. The percentage of singlet oxygen scavenging was calculated from the equation

$$\% \text{ Singlet oxygen scavenging} = [A_{control} - (A_{sample} - A_{sample\ blank})/A_{control}] \times 100, \quad (4)$$

where $A_{control}$, A_{sample}, and $A_{sample\ blank}$ are the absorbance of the mixture solution, astaxanthin with mixture solution and astaxanthin without mixture solution, respectively.

2.5. Determination of Tyrosinase Inhibitory Activity

Tyrosinase inhibitory activity was measured according to the modified dopachrome method [21]. In brief, 50 mL of astaxanthin extract in ethanol at various concentrations (3–50 μg/mL) was applied to a 96-well plate followed by the addition of 50 μL of 20 mM phosphate buffer (pH 6.8) and mushroom tyrosinase (500 units/mL). Each well was mixed and incubated at room temperature for 15 min. Then, 50 μL of 0.85 mM L-3,4-dihydroxyphenylalanine (L-DOPA) was added. After incubation at room temperature for 10 min, the mixture was measured for the difference of the absorbance at 492 nm before and after incubation. Kojic acid and arbutin were used as positive controls and the values were expressed as 50% inhibitory concentration (IC_{50}). The percentage of tyrosinase inhibition was calculated as

$$\%\text{inhibition} = [(A-B) - (C-D)]/(A-B) \times 100, \quad (5)$$

where A is the absorbance of the blank after incubation, B is the absorbance of the blank before incubation, C is the absorbance of the sample after incubation, and D is the absorbance of the sample before incubation.

2.6. Cytotoxicity Assay

2.6.1. Cell Culture

Human dermal fibroblasts (WS1) were purchased from American Type Culture Collection (ATCC, Manassas, VA, USA). The cells were cultured in Dulbecco's modified Eagle's medium (DMEM) supplemented with 10% fetal bovine serum (FBS) and 1% L-glutamine and 1% penicillin-streptomycin in 5% CO_2 environment at 37 °C until they reached a confluence of 70–80%.

2.6.2. Cell Viability

Cell viability was determined using an MTT assay as described by previous report [22]. The cells were seeded at 1×10^5 cells/well in a 96-well plate and were treated with 100 µL of astaxanthin in at various concentrations (5–160 µg/mL), and then the plates were incubated at 37 °C with 5% CO_2 for 24 h. The test compound was dissolved in dimethyl sulfoxide (DMSO), and the final concentration of DMSO in the medium was 1% *v/v*. For the control, no astaxanthin was added (none). Thereafter, 100 µL of MTT solution was added to each well and incubation performed for 2 h. After 2 h, the formazan crystals were dissolved by adding 100 µL DMSO and the plate was further incubated for 5 min at room temperature. The absorbance was measured at 570 nm using a microplate reader. Percentage of cell viability was calculated by comparing absorbance values of samples with those of the control.

2.6.3. Microscopic Observations

The changes in morphology and detachment of human dermal fibroblasts were observed after incubation with astaxanthin using a Nikon inverted microscope (Nikon Eclipse TS100, Nikon, Japan) equipped with an objective lens (Plan 10/0.25DL/Ph1, Nikon, Japan) of ×10 magnification.

2.7. Statistical Analysis

All results are expressed as mean ± SD (standard deviation). Analysis of variance was performed, and significant differences between means were determined by Duncan's multiple range tests at a level of $p < 0.05$.

3. Results and Discussion

3.1. Yield and Levels of Astaxanthin

The ethanol extracted astaxanthin yield in the form of red-orange paste was 28.9 ± 3.2 mg/g shrimp shells. This result is similar to a previous study by Taksima et al. [4] who reported that extraction of astaxanthin from shrimp shells with ethanol obtained a yield of 24.7 ± 2.9 mg/g shrimp shells. The chromatographic condition used in this study gave a good resolution for the analysis of astaxanthin. The retention time of astaxanthin was 5.3 min which was found to be close to those of the astaxanthin standard. It was found that astaxanthin was the major component of shrimp extract representing about 45% among the extract (15.6 ± 0.6 mg astaxanthin/g extract). Some researchers have revealed that astaxanthin was the major carotenoid pigment with a relative percentage about 65–98% of the total carotenoid content, whereas free astaxanthin accounted for a large percentage [14,23]. For organic solvent and alcoholic extraction of astaxanthin from shrimp shells, some authors [3,5] have reported values of extract 47.86 µg/g extract and 9.27 mg/g, respectively. Thus, a higher content of astaxanthin was observed for the shrimp shells used in the present study.

3.2. Antioxidant Properties of Astaxanthin Extract

3.2.1. DPPH Radical Scavenging Activity

The DPPH radical scavenging assay has been commonly used to evaluate the free radical scavenging activities of antioxidants. Due to the hydrogen donating ability of antioxidants, they may reduce the free radical DPPH• to a stable form, with a decrease in absorbance at 517 nm [24]. Based on this principle, various concentrations of astaxanthin were measured for their DPPH radical scavenging activities, and the results are presented in Table 1. All samples at tested concentrations showed obvious scavenging activities on DPPH radicals in a dose-dependent manner. The lower EC_{50} value means the more powerful antioxidant capacity. Ethanol extract of astaxanthin presented antioxidant potential with the EC_{50} value of 17.5 ± 3.6 µg/mL at which the astaxanthin was comparable to BHT (17.2 ± 0.1 µg/mL). The DPPH radical scavenging activity of astaxanthin was also higher than that of the positive control, ascorbic acid (EC_{50}: 19.7 ± 0.2 µg/mL). This result indicated that astaxanthin from shrimp shells had a potent antioxidant property due to its unique molecular structure, which consists of hydrocarbon with conjugated double bonds called poliene, whereas the presence of the hydroxyl (OH) and keto (C=O) endings on each ionone ring resulting in astaxanthin has hydrogen donating capability [25]. The result was consistent with Sowmya and Sachindra [23] who reported that astaxanthin from shrimp waste (*Penaeus indicus*) showed strong antioxidant activity as indicated by radical scavenging, comparable to that of the known antioxidant α-tocopherol.

Table 1. Antioxidant properties of astaxanthin extract as measured by different antioxidant assays

Samples	Antioxidant Activities (EC$_{50}$, µg/mL)			
	DPPH	**ABTS**	**β-Carotene Bleaching**	**Singlet Oxygen**
Astaxanthin	17.5 ± 3.6 [b]	7.7 ± 0.6 [c]	15.1 ± 1.9 [a]	9.2 ± 0.5 [c]
Ascorbic acid	19.7 ± 0.2 [a]	20.8 ± 1.1 [a]	12.5 ± 0.3 [b]	-
BHT	17.2 ± 0.1 [b]	15.1 ± 0.7 [b]	11.5 ± 0.1 [b]	-
Rutin	-	-	-	55.0 ± 1.6 [a]
Quercetin	-	-	-	50.5 ± 1.1 [b]

[a–c] Mean values in the same column followed by the different letters are significantly different ($p < 0.05$); BHT: Butylated hydroxytoluene.

3.2.2. ABTS Radical Scavenging Activity

ABTS radical scavenging assay is based on the ability of antioxidants to reduce the preformed radical ABTS with its consequent decolorization of a blue-green color in the absorbance at 734 nm [26]. In the ABTS assay, scavenging trends showed similar patterns of DPPH assay results. The astaxanthin from shrimp shells showed a concentration-dependent ABTS scavenging activity with EC$_{50}$ value of 7.7 ± 0.6 µg/mL (Table 1). Apparently, the astaxanthin had significantly ABTS scavenging activity and more potent than ascorbic acid and BHT (EC$_{50}$: 20.8 ± 1.1 and 15.1 ± 0.7 µg/mL, respectively). This result confirmed that astaxanthin had noticeable effect on scavenging ABTS radicals.

3.2.3. β-Carotene Bleaching (BCB) Activity

The BCB method used to evaluate the ability of a compound to prevent β-carotene oxidation by protecting it against free radicals generated during linoleic acid peroxidation. A model system undergoes discoloration in the absence of an antioxidant and it may be oxidized and subsequently, the system loses its orange color [27]. Table 1 shows the antioxidant activity of ethanol extract of astaxanthin from shrimp shells as measured by the bleaching of β-carotene. The result demonstrates a similar trend to DPPH and ABTS radical scavenging activities, where the BCB antioxidant activity increased with increasing concentration of the extract used. The EC$_{50}$ value was found to be 15.1 ± 1.9 µg/mL. The results indicated that astaxanthin could hinder the extent of β-carotene bleaching by neutralizing the linoleate-free radical and other free radicals formed in the system. However, it was less active than the positive controls, ascorbic acid, and BHT (EC$_{50}$: 12.5 ± 0.3 and 11.5 ± 0.1 µg/mL, respectively).

3.2.4. Singlet Oxygen Scavenging ($^1O_2$) Activity

Singlet oxygen ($^1O_2$) is a non-radical ROS with one of the strongest activities. It is involved in many diseases and skin disorders. Hence, the singlet oxygen scavenging activity of natural compounds indicates their usefulness as antioxidants. The quenching of singlet oxygen of astaxanthin extract, rutin, and quercetin from the present study are shown in Table 1. Astaxanthin extract exhibited strong singlet oxygen quenching activity with EC$_{50}$ value of 9.2 ± 0.5 µg/mL. Interestingly, the extract had significantly strong ability of ROS inhibition as compared with rutin and quercetin. This result is supported by Nishida et al. [28] who reported that astaxanthin is known to have potent singlet oxygen quenching activity and the efficiency of singlet oxygen quenching of the carotenoids has been shown to be related to the number of conjugated double bonds. As with the astaxanthin extract from shrimp shells used in the study, it exhibits strong singlet oxygen quenching activity, which indicates its usefulness as a natural antioxidant for skin and disease prevention.

3.3. Tyrosinase Inhibition

Tyrosinase is an enzyme that is involved in the rate limiting step for the control of melanin production. Therefore, the inhibition of tyrosinase activity leads to induce skin whitening due to a reduction of melanin synthesis. When the tyrosinase enzyme was incubated with the astaxanthin extract, it inhibited tyrosinase activity in a dose-dependent manner at concentrations of 3–50 µg/mL. The results demonstrated that astaxanthin exhibited potent inhibitory effects against mushroom tyrosinase. The IC_{50} value of astaxanthin extracts against tyrosinase was 12.2 ± 1.5 µg/mL (Figure 1). The astaxanthin indicated relatively same inhibitory activity compared with positive control antioxidant, arbutin (IC_{50}: 10.1 ± 0.2 µg/mL), however show less antityrosinase activity than kojic acid (IC_{50}: 7.7 ± 0.5 µg/mL). The presence of two oxygenated groups on each ring structure, which is like a phenolic compound of astaxanthin, may be responsible for the inhibition of tyrosinase enzyme due to its ability to chelate copper in the active site [29]. This result is in concordance with the report from Xue et al. [30], that phenolic compounds such as kaempferol and quercetin were able to chelate copper atoms of tyrosinase enzyme and further inhibited tyrosinase activity. Moreover, this study showed that shrimp astaxanthin extract possessed higher tyrosinase inhibitory effect than phenolic compounds from mushroom extracts [31] and marigold flower extracts [32].

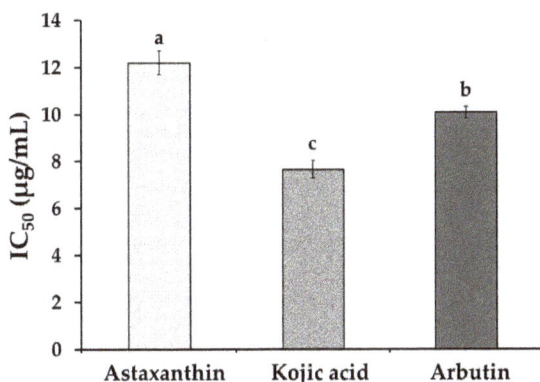

Figure 1. Tyrosinase inhibition activity of astaxanthin extract, kojic acid, and arbutin. Values are expressed as 50% inhibitory concentration (IC_{50}). Letters with different superscripts indicate samples that are significantly different ($p < 0.05$) from each other.

3.4. Cytotoxicity Assay

3.4.1. Cell Viability

The in vitro cytotoxicity assay demonstrated obvious correlation to the toxicity in vivo [33]. MTT assay is a colorimetric method that measures the reduction of yellow tetrazolium bromide into an insoluble purple formazan product by mitochondrial succinate dehydrogenase. The amount of dye converted is directly proportional to the number of living cells [34], so it is a direct reflection of cytotoxicity. In this study, human dermal fibroblast cells (WS1) were used for toxicity evaluation. The results of cytotoxic activity of astaxanthin extract from shrimp shells are summarized in Figure 2. The viability of the cells treated with a concentration of astaxanthin in the range of 5-160 µg/mL for 24 h were not significantly different from the control, even tends to decrease with the increase in concentration. Obviously, the astaxanthin extract at all concentrations was non-toxic to human dermal fibroblast cells and gave high level of cell viability (more than 90%); thus, the half inhibition concentration (IC_{50}) could not be determined. This implies that the extract does not affect the viability of the studied human skin cells. Therefore, within a reasonable dosage range, the application of astaxanthin should be safe for cosmetic development.

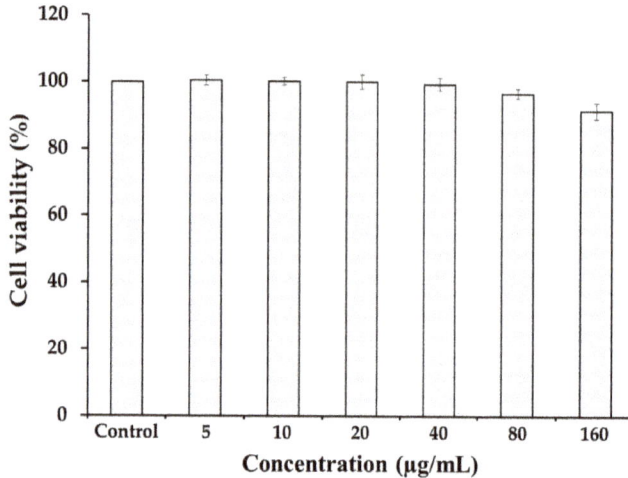

Figure 2. Cell viability of human dermal fibroblast cell lines exposed to astaxanthin at different concentrations for 24 h determined by 3-(4,5-dimethylthiazol-2-yl)-2,5-diphenyltetrazolium bromide (MTT) assay.

3.4.2. Cell Morphological Characteristics

To confirm astaxanthin induced cytotoxicity, we investigated the changes of cell morphology in WS1 human dermal fibroblast cells exposed to 5–160 µg/mL astaxanthin for 24 h. Images of cells under magnification are shown in Figure 3. Human dermal fibroblast cells are typically flattened or extensible shaped, adherent cells growing as a confluent monolayer (Figure 3A). For cells treated with a concentration of astaxanthin in the range of 5–160 µg/mL, no change of cell morphology could be detected compared with control cells (Figure 3B–G). Similar results were observed by other studies of fibroblast cells treated with natural extracts. Itsarasook et al. [35] reported that *Terminalia Chebula* fruit extract showed no cytoxic effect to human skin fibroblasts even at the highest concentration of 50 µg/mL because the viability of the studied cells was not affected either in the number of growth or the morphology. In this study, astaxanthin extract at all concentrations was non-toxic to human dermal fibroblasts, as shown by the cell viability higher than 90%. Meanwhile, cells induced by ascorbic acid, (the positive control) generated more pronounced cell debris and changes in morphology such as shrinkage, roundness, and detachment from the surface (Figure 3H). These results confirmed that astaxanthin from shrimp shells displayed a non-toxic effect and might be an alternative ingredient for use in cosmetics.

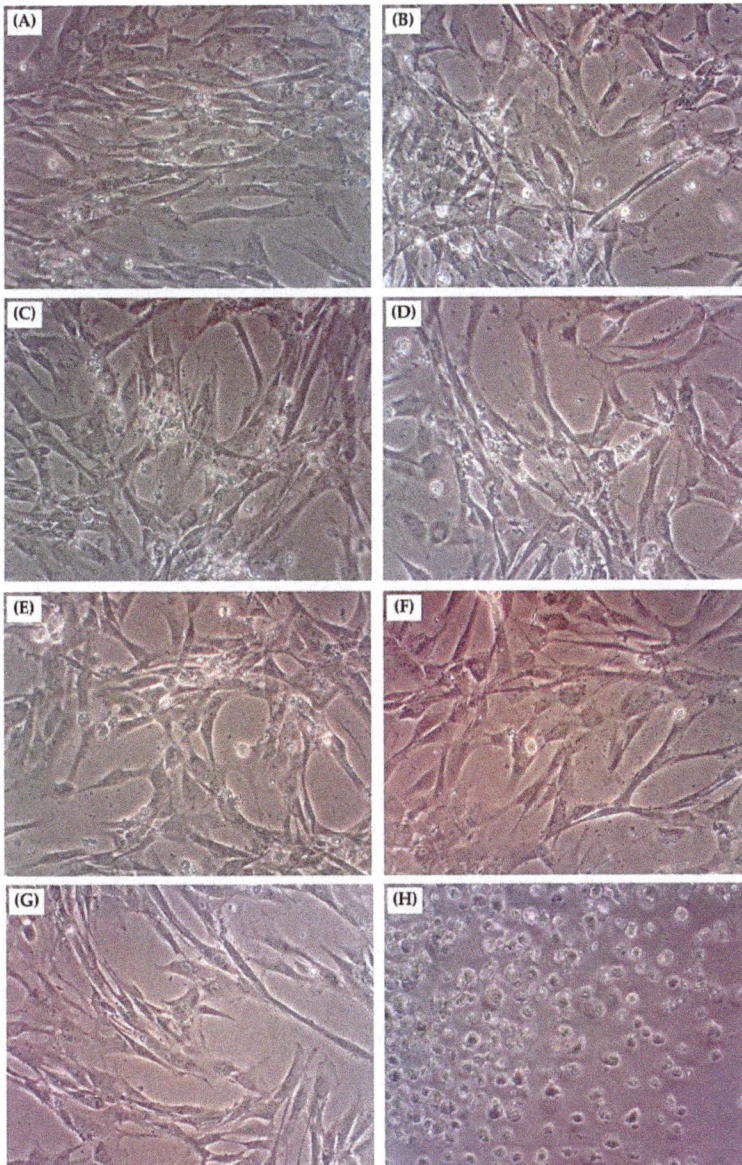

Figure 3. Morphological changes of human dermal fibroblast cells after exposure to astaxanthin extract for 24 h (magnification ×10). (**A**) Control cells, (**B**) cells treated with 5 µg/mL astaxanthin, (**C**) 10 µg/mL astaxanthin, (**D**) 20 µg/mL astaxanthin, (**E**) 40 µg/mL astaxanthin, (**F**) 80 µg/mL astaxanthin, (**G**) 160 µg/mL astaxanthin, and (**H**) cells treated with 100 µg/mL ascorbic acid (positive control).

4. Conclusions

This study can be considered as the first attempt to evaluate astaxanthin from shrimp shells for possible antioxidant, antityrosinase, and cytotoxicity properties. Results from this study showed that astaxanthin possesses potent antioxidant activity and tyrosinase inhibition. Remarkably, the astaxanthin was non-cytotoxic towards human dermal fibroblast cells. We suggest that concentrations of astaxanthin

Antioxidants **2019**, *8*, 128

in the range of 10–20 µg/mL exhibited optimal levels of antioxidant and antityrosinase activities. In addition, astaxanthin concentrations up to 160 µg/mL had no cytotoxic effects. Our findings confirmed that shrimp astaxanthin can be used as a functional ingredient in skin health products with both the efficacy and safety.

Author Contributions: Conceptualization, S.C. and W.K.; Methodology, S.C., W.P., and U.R.-A.; Software, S.C.; Validation, W.P., U.R.-A., Y.W., and W.K.; Formal analysis, S.C.; Investigation, S.C.; Resources, W.P., U.R.-A., and W.K.; Data curation, S.C. and W.K.; Writing—original draft preparation, S.C.; Writing—review and editing, W.K. and Y.W.; Visualization, S.C.; Supervision, W.P.; Project administration, W.P.; Funding acquisition, W.P. and U.R.-A.

Funding: Thailand Institute of Scientific and Technological Research (Funding number: 6016301010).

Acknowledgments: The authors thank Thailand Institute of Scientific and Technological Research for the use of research facilities.

Conflicts of Interest: The authors declare no conflict of interest.

References

1. Naguib, Y.M. Antioxidant activities of astaxanthin and related carotenoids. *J. Agric. Food Chem.* **2000**, *48*, 1150–1154. [CrossRef] [PubMed]
2. Fakhri, S.; Abbaszadeh, F.; Dargahi, L.; Jorjani, M. Astaxanthin: A mechanistic review on its biological activities and health benefits. *Pharmacol. Res.* **2018**, *136*, 1–20. [CrossRef] [PubMed]
3. Sachindra, N.M.; Bhaskar, N.; Siddegowda, G.S.; Sathisha, A.D.; Suresh, P.V. Recovery of carotenoids from ensilaged shrimp waste. *Bioresour. Technol.* **2007**, *98*, 1642–1646. [CrossRef]
4. Taksima, T.; Limpawattana, M.; Klaypradit, W. Astaxanthin encapsulated in beads using ultrasonic atomizer and application in yogurt as evaluated by consumer sensory profile. *LWT Food Sci. Technol.* **2015**, *62*, 431–437. [CrossRef]
5. Santos, S.D.; Cahú, T.B.; Firmino, G.O.; de Castro, C.C.M.M.B.; Carvalho, J.L.B.; Bezerra, R.S.; Filho, J.L.L. Shrimp waste extract and astaxathin: Rat alveolar macrophage, oxidative stress and inflammation. *J. Food Sci.* **2012**, *77*, 141–146. [CrossRef] [PubMed]
6. Tominaga, K.; Hongo, N.; Karato, M.; Yamashita, E. Protective effects of astaxanthin against singlet oxygen induced damage in human dermal fibroblasts in vitro. *Food Style 21* **2009**, *13*, 84–86.
7. Seki, T.; Sueki, H.; Kono, H.; Suganuma, K.; Yamashita, E. Effects of astaxanthin from *Haematococcus pluvialis* on human skin-patch test; skin repeated application test; effect on wrinkle reduction. *Fragr. J.* **2001**, *12*, 98–103.
8. Tominaga, K.; Hongo, N.; Karato, M.; Yamashita, E. Cosmetic benefits of astaxanthin on humans subjects. *Acta Biochim. Pol.* **2012**, *59*, 43–47. [CrossRef] [PubMed]
9. Yamashita, E. Suppression of post-UVB hyperpigmentation by topical astaxanthin from krill. *Fragr. J.* **2001**, *12*, 98–103.
10. Arakane, K. Superior skin protection via astaxanthin. *Carotenoid Sci.* **2002**, *5*, 21–24.
11. Nakajima, H.; Fukazawa, K.; Wakabayashi, Y.; Wakamatsu, K.; Senda, K.; Imokawa, G. Abrogating effect of a xanthophyll carotenoid astaxanthin on the stem cell factor-induced stimulation of human epidermal pigmentation. *Arch. Dermatol. Res.* **2012**, *304*, 803–816. [CrossRef]
12. Rao, A.R.; Sindhuja, H.N.; Dharmesh, S.M.; Sankar, K.U.; Sarada, R.; Ravishankar, G.A. Effective inhibition of skin cancer, tyrosinase, and antioxidative properties by astaxanthin and astaxanthin esters from the green alga *Haematococcus pluvialis*. *J. Agric. Food Chem.* **2013**, *61*, 3842–3851. [CrossRef]
13. Tominaga, K.; Hongo, N.; Fujishita, M.; Takahashi, Y.; Adachi, Y. Protective effects of astaxanthin on skin deterioration. *J. Clin. Biochem. Nutr.* **2017**, *61*, 33–39. [CrossRef]
14. Prameela, K.; Venkatesh, K.; Immandi, S.B.; Kasturi, A.P.K.; Krishna, C.R.; Mohan, C.M. Next generation nutraceutical from shrimp waste: The convergence of applications with extraction methods. *Food Chem.* **2017**, *237*, 121–132. [CrossRef]
15. Shah, M.M.; Liang, Y.; Cheng, J.J.; Daroch, M. Astaxanthin-producing green microalga *Haematococcus pluvialis*: From single cell to high value commercial products. *Front. Plant Sci.* **2016**, *7*, 531. [CrossRef]
16. Higuera-Ciapara, I.; Félix-Valenzuela, L.; Goycoolea, F.M. Microencapsulation of astaxanthin in a chitosan matrix. *Carbohydr. Polym.* **2004**, *56*, 41–45. [CrossRef]

17. Navarro-Hoyos, M.; Alvarado-Corella, D.; Moreira-Gonzalez, I.; Arnaez-Serrano, E.; Monagas-Juan, M. Polyphenolic composition and antioxidant activity of aqueous and ethanolic extracts from *Uncaria tomentosa* bark and leaves. *Antioxidants* **2018**, *7*, 65. [CrossRef]

18. Daniela, S.; Bubelova, Z.; Sneyd, J.; Erb, S.; Mleck, J. Total phenolic, flavonoids, antioxidant capacity, crude fiber and digestibility in non-traditional wheat flakes and muesli. *Food Chem.* **2015**, *174*, 319–325.

19. Gouveia-Figueira, S.C.; Gouveia, C.A.; Carvalho, M.J.; Rodrigues, A.I.; Nording, M.L.; Castilho, P.C. Antioxidant capacity, cytotoxicity and antimycobacterial activity of madeira archipelago endemic *Helichrysum* dietary and medicinal plants. *Antioxidants* **2014**, *3*, 713–729. [CrossRef]

20. Tu, P.T.B.; Tawata, S. Anti-oxidant, anti-aging, and anti-melanogenic properties of the essential oils from two varieties of *Alpinia zerumbet*. *Molecules* **2015**, *20*, 16723–16740. [CrossRef]

21. Tadtong, S.; Viriyaroj, A.; Vorarat, S.; Nimkulrat, S.; Suksamrarn, S. Antityrosinase and antibacterial activities of mangosteen pericarp extract. *J. Health Res.* **2009**, *23*, 99–102.

22. Silva, A.R.; Seidl, C.; Furusho, A.S.; Boeno, M.M.S.; Dieamant, G.C.; Weffort-Santos, A.M. In vitro evaluation of the efficacy of commercial green tea extracts in UV protection. *Inter. J. Cosmet. Sci.* **2013**, *35*, 69–77. [CrossRef]

23. Sowmya, R.; Sachindra, N.M. Evaluation of antioxidant activity of carotenoid extract from shrimp processing by products by in vitro assays and in membrane model system. *Food Chem.* **2012**, *134*, 308–314. [CrossRef]

24. Craft, B.D.; Kerrihard, A.L.; Amarowicz, R.; Pegg, R.B. Phenol-based antioxidants and the in vitro methods used for their assessment. *Compr. Rev. Food Sci. Food Saf.* **2012**, *11*, 149–173. [CrossRef]

25. Liu, X.; Osawa, T. Cis astaxanthin and especially 9-cis astaxanthin exhibits a higher antioxidant activity in vitro compared to the all-trans isomer. *Biochem. Biophys. Res. Commun.* **2007**, *357*, 187–193. [CrossRef]

26. Prior, R.L.; Wu, X.; Schaich, K. Standardized methods for the determination of antioxidant capacity and phenolics in foods and dietary supplements. *J. Agric. Food Chem.* **2005**, *53*, 4290–4302. [CrossRef]

27. Jayaprakasha, G.K.; Singh, R.P.; Sakariah, K.K. Antioxidant activity of grape seed (*Vitis vinifera*) extracts on peroxidation models in vitro. *Food Chem.* **2001**, *73*, 285–290. [CrossRef]

28. Nishida, Y.; Yamashita, E.; Miki, W. Quenching activities of common hydrophilic and lipophilic antioxidants against singlet oxygen using chemiluminescence detection system. *Carotenoid Sci.* **2007**, *11*, 16–20.

29. Loizzo, M.R.; Tundis, R.; Menichini, F. Natural and synthetic tyrosinase inhibitors as antibrowning agents: An update. *Compr. Rev. Food Sci. Food Saf.* **2012**, *11*, 378–398. [CrossRef]

30. Xue, Y.L.; Miyakawa, T.; Hayashi, Y.; Okamoto, K.; Hu, F.; Mitani, N.; Furihata, K.; Sawano, Y.; Tanokura, M. Isolation and tyrosinase inhibitory effects of polyphenols from the leaves of Persimmon, Diospyros kaki. *J. Agric. Food Chem.* **2011**, *59*, 6011–6017. [CrossRef]

31. Banlangsawan, N.; Sripanidkulchai, B.; Sanoamuang, N. Investigation of antioxidative, antityrosinase and cytotoxic effects of extract of irradiated oyster mushroom. *Songklanakarin J. Sci. Technol.* **2016**, *38*, 31–39.

32. Vallisuta, O.; Nukoolkarn, V.; Mitrevej, A.; Sarisuta, N.; Leelapornpisid, P.; Phrutivorapongkul, A.; Sinchaipanid, N. In vitro studies on the cytotoxicity, and elastase and tyrosinase inhibitory activities of marigold (*Tagetes erecta* L.) flower extracts. *Exp. Ther. Med.* **2014**, *7*, 246–250. [CrossRef]

33. Cheong, H.I.; Johnson, J.; Cormier, M.; Hosseini, K. In vitro cytotoxicity of eight beta-blockers in human corneal epithelial and retinal pigment epithelial cell lines: Comparison with epidermal keratinocytes and dermal fibroblasts. *Toxicol. In Vitro* **2008**, *22*, 1070–1076. [CrossRef]

34. Han, J.; Back, S.H.; Hur, J.; Lin, Y.H.; Gildersleeve, R.; Shan, J. ER-stress-induced transcriptional regulation increases protein synthesis leading to cell death. *Nat. Cell Biol.* **2013**, *15*, 481–490. [CrossRef]

35. Itsarasook, K.; Ingkaninan, K.; Chulasiri, M.; Viyoch, J. Antioxidant activity and cytotoxicity to human skin fibroblasts of *Terminalia Chebula* fruit extract. In Proceedings of the 2012 International Conference on Biological and Life Sciences, Singapore, 23–24 July 2012; IACSIT Press: Singapore, 2012.

antioxidants

MDPI

Article

First Apocarotenoids Profiling of Four Microalgae Strains

Mariosimone Zoccali [1], Daniele Giuffrida [2,*], Fabio Salafia [1], Carmen Socaciu [3], Kari Skjånes [4], Paola Dugo [1,5,6,7] and Luigi Mondello [1,5,6,7]

[1] Department of Chemical, Biological, Pharmaceutical and Environmental Sciences, University of Messina, 98166 Messina, Italy
[2] Department of Biomedical, Dental, Morphological and Functional Imaging Sciences, University of Messina, 98125 Messina, Italy
[3] PROPLANTA-Research Centre for Applied Biotechnology, str. Trifoiului 12G, 400478 Cluj-Napoca, Romania
[4] Division of Biotechnology and Plant Health, The Norwegian Institute of Bioeconomy Research (NIBIO), PO115, N-1431 Ås, Norway
[5] Chromaleont s.r.l., c/o Department of Chemical, Biological, Pharmaceutical and Environmental Sciences, University of Messina, 98166 Messina, Italy
[6] BeSep s.r.l., c/o Department of Chemical, Biological, Pharmaceutical and Environmental Sciences, University of Messina, 98166 Messina, Italy
[7] Unit of Food Science and Nutrition, Department of Medicine, University Campus Bio-Medico of Rome, 00128 Rome, Italy
* Correspondence: dgiuffrida@unime.it; Tel.: +39-090-676-6996

Received: 18 June 2019; Accepted: 3 July 2019; Published: 6 July 2019

Abstract: Both enzymatic or oxidative carotenoids cleavages can often occur in nature and produce a wide range of bioactive apocarotenoids. Considering that no detailed information is available in the literature regarding the occurrence of apocarotenoids in microalgae species, the aim of this study was to study the extraction and characterization of apocarotenoids in four different microalgae strains: *Chlamydomonas* sp. CCMP 2294, *Tetraselmis chuii* SAG 8-6, *Nannochloropsis gaditana* CCMP 526, and *Chlorella sorokiniana* NIVA-CHL 176. This was done for the first time using an online method coupling supercritical fluid extraction and supercritical fluid chromatography tandem mass spectrometry. A total of 29 different apocarotenoids, including various apocarotenoid fatty acid esters, were detected: apo-12'-zeaxanthinal, β-apo-12'-carotenal, apo-12-luteinal, and apo-12'-violaxanthal. These were detected in all the investigated strains together with the two apocarotenoid esters, apo-10'-zeaxanthinal-C4:0 and apo-8'-zeaxanthinal-C8:0. The overall extraction and detection time for the apocarotenoids was less than 10 min, including apocarotenoids esters, with an overall analysis time of less than 20 min. Moreover, preliminary quantitative data showed that the β-apo-8'-carotenal content was around 0.8% and 2.4% of the parent carotenoid, in the *C. sorokiniana* and *T. chuii* strains, respectively. This methodology could be applied as a selective and efficient method for the apocarotenoids detection.

Keywords: carotenoid derivatives; microphytes; supercritical fluid extraction-supercritical fluid chromatography-tandem mass spectrometry; hyphenated techniques

---

1. Introduction

The carotenoids composition of microalgae has been widely investigated [1–4] and, recently, the occurrence of carotenoids esters in microalgae has also been reported [5]. The carotenoid profiles are known to vary greatly between species, as are the algae's ability to accumulate different carotenoids during stress exposure [6]. The production of carotenoids from microalgae is continuously growing since natural and controlled production sources of carotenoids are highly desirable because of their economic and environmental positive aspects [7]. Carotenoids are tetraterpenoidic lipophilic compounds with health beneficial properties, such as antioxidant activity [8,9], composed of two main classes: the carotenes that are hydrocarbons molecules and the xanthophylls that are oxygenated ones. It is very common in nature to find xanthophylls esterified with fatty acids; in fact, xanthophyll esters have greater stability then free xanthophylls. Different analytical methods for extraction and analysis of carotenoids in microalgae samples were reported mainly based on liquid extraction and liquid chromatography approaches [10], but they were also based on supercritical fluids approaches [11,12]. Both enzymatic or oxidative carotenoids cleavages often occur in plants that produce a wide range of bioactive apocarotenoids [13,14]. Possible zeaxanthin oxidative cleavage sites that produce various apozeaxanthinals are shown in Figure 1. There is a growing interest in the investigation of apocarotenoids in food, food products, and mammals due to the beneficial effects attributed to them [14–16]. Very recently, Zoccali et al. [17] and Giuffrida et al. [18] reported on the first application of a supercritical fluid extraction-supercritical fluid chromatography-mass spectrometry (SFE-SFC-MS) methodology for, respectively, the carotenoids and the apocarotenoids determination in different food matrices. To the best of the authors knowledge, no detailed data is available in the literature on the apocarotenoids occurrence in microalgae. Therefore, the aim of this investigation was to determine the occurrence of apocarotenoids in four selected different microalgae strains: *Chlamydomonas* sp CCMP 2294, *Tetraselmis chuii* SAG 8-6, *Nannochloropsis gaditana* CCMP 526, and *Chlorella sorokiniana* NIVA-CHL 176.

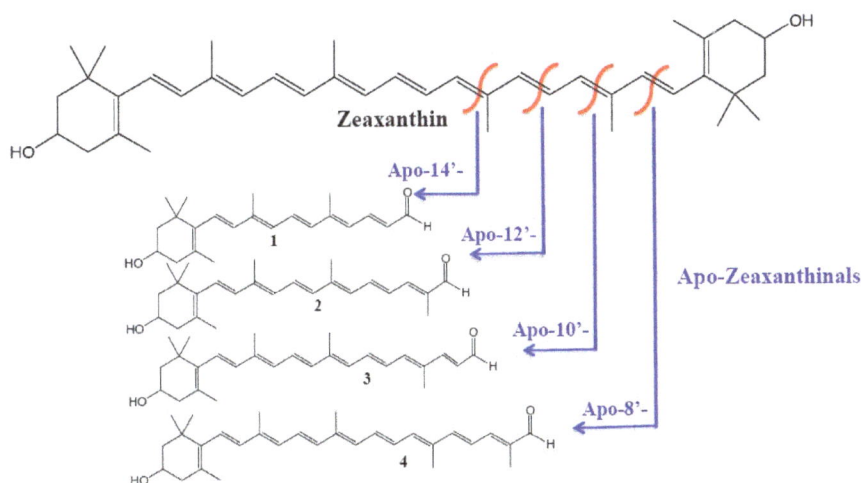

Figure 1. Zeaxanthin oxidative cleavages sites producing various apozeaxanthinals; **1**. Apo-14'-Zeaxanthinal; **2**. Apo-12'-Zeaxanthinal; **3**. Apo-10'-Zeaxanthinal; **4**. Apo-8'-Zeaxanthinal. Reprint with permission from [14].

2. Materials and Methods

2.1. Chemicals

All chemicals were obtained from Merck Life Science (Merck KGaA, Darmstadt, Germany). A series of β-apocarotenals, apozeaxanthinals, and ε-apoluteinals were generated by oxidative cleavages of the parent carotenoids as reported in references [19–21]; moreover, the β-apo-8'-Carotenal standard was purchased from CaroteNature GmbH (Münsingen, Switzerland). The standards of the parent carotenoids, namely, β-carotene, zeaxanthin, and lutein were obtained from Extrasynthese (Genay, France).

2.2. Strain Selection and Biomass Production

The following four different microalgae strains were acquired from culture collections and were thusly investigated:

Chlamydomonas sp CCMP 2294 was obtained from Bigelow Laboratory for Ocean Sciences (NCMA), USA.

Artic marine collection site: Baffin Bay, between Ellesmere Island, Canada and Greenland (77.8136° N 76.3697° W, sea ice core), belonging to the Chlamydomonadaceae family.

Cultivated under the following conditions: Light intensity: 70–80 μmol/m^2/s, temperature 4 °C, 6 L cultures in 10 L flasks bubbled with air added 1% CO_2, growth medium L1 [22].

Tetraselmis chuii SAG 8-6 was obtained from SAG Culture Collection of Algae, Germany.

Temperate marine collection site: Scotland, Millport, Clyde estuary (55.751383/–4.931953, 600 m), belonging to the Chlorodendraceae family.

Cultivated under the following conditions: Light intensity: 130 μmol/m^2/s, temperature 25 °C, 1 L cultures in 2 L Erlenmeyer flasks on shaking table, air with 3% CO_2 was added to headspace; Light intensity: 50 μmol/m^2/s, temperature 22 °C, 5–6 L cultures in 10 L flasks bubbled with air added 3% CO_2, both with growth medium L1.

Nannochloropsis gaditana CCMP 526 (recently renamed after full genome completion, as *Microchloropsis gaditana*) was obtained from Bigelow Laboratory for Ocean Sciences (NCMA), USA.

Temperate marine collection site: Morocco, Lagune de Oualidia, (32.8333° N 9° W), belonging to the Eustigmataceae family.

Cultivated under the following conditions: Light intensity: 130 μmol/m^2/s, temperature 25 °C, 1 L cultures in 2 L Erlenmeyer flasks on shaking table, air with 3% CO_2 was added to headspace; Light intensity: 50 μmol/m^2/s, temperature 22 °C, 5–6L cultures in 10 L flasks bubbled with air added 3% CO_2, both with growth medium L1.

Chlorella sorokiniana NIVA-CHL 176 was obtained from The Norwegian Culture Collection of Algae (NORCCA), Norway.

Temperate fresh water collection site: Waller Creek, University of Texas, Austin, USA, belonging to the Chlorellaceae family.

Cultivated under the following conditions: Light intensity: 150 μmol/m^2/s, temperature 25 °C, 1 L cultures in 1,2 L flat flasks bubbled with air added 2–3% CO_2. Growth medium Tris-Acetate-Phosphate (TAP) [23], modified by replacing acetate with HCl.

All the above described microalgae biomasses were lyophilized before apocarotenoids analyses.

2.3. Sample Preparation

The microalgae samples (1 mg) were placed in the extraction vessel in the SFE unit. A 0.2 mL extraction vessel was used. Supercritical CO_2 and CH_3OH were then utilized to perform the extraction and then the chromatography as reported in Section 2.5.

2.4. SFE-SFC-MS Instrumentation

The SFE-SFC-MS analyses were carried out on a Shimadzu Nexera-UC system (Shimadzu, Kyoto, Japan), composed of a CBM-20A controller, an SFE-30A module for supercritical fluid extraction, two LC-20AD$_{XR}$ dual-plunger parallel-flow pumps, an LC-30AD$_{SF}$ CO_2 pump, two SFC-30A back pressure regulator, a DGU degasser, a CTO-20AC column oven, a SIL-30AC autosampler, an LCMS-8050 mass spectrometer equipped with an atmospheric pressure chemical ionization (APCI) source. The all system was controlled by the LabSolution ver. 5.8 (Shimadzu, Kyoto, Japan).

2.5. SFE-SFC-MS Analytical Conditions

A scheme of the SFE-SFC-MS system is reported in Figure 2 and described in detail in Zoccali et al. [17]. The system operates in three different steps: (1) SFE static extraction mode, (2) SFE dynamic extraction mode, and (3) SFC analysis. During the static extraction mode, the vessel was pressurized for 3 min (Figure 2A), then the extraction was carried out in the dynamic mode for one min (Figure 2B). During this step, the mobile phase flows through the vessel continuously and the extracts are transferred into the analytical column. After the SFEs steps 1 and 2, the analytes undergo the SFC analysis (Figure 2C).

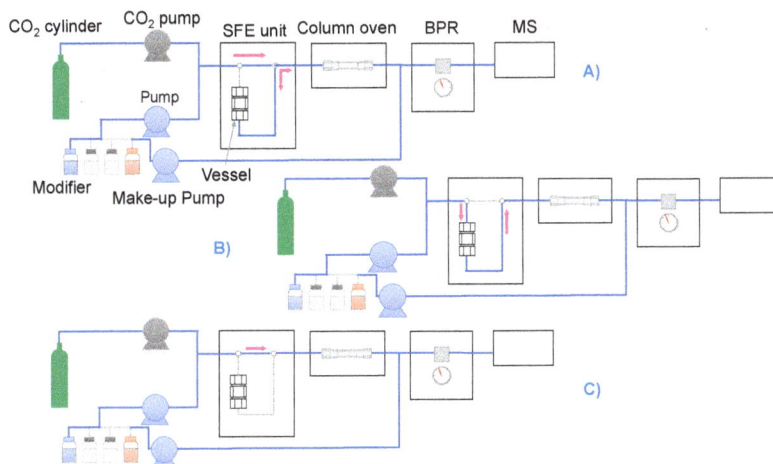

Figure 2. Scheme of the supercritical fluid extraction-supercritical fluid chromatography-mass spectrometry (SFE-SFC-MS) system: (**A**) Static extraction mode, (**B**) Dynamic extraction mode, (**C**) Analysis mode. Reprinted with permission from [17].

The SFE conditions were as follows: 0–3 min static extraction mode, and 3–4 min dynamic extraction mode; Extraction vessel temperature: 80 °C. Back pressure regulator: 150 bar.

Solvent (A) CO_2 and solvent (B) CH_3OH; Gradient: From 0 to 3 min, 5% of B; then from 3 to 4 min, 10% of B. Flow rate: 2 mL/min.

The SFC conditions were as follows: Solvent (A) CO_2 and solvent (B) CH_3OH. Gradient: from 4 to 6.0 min 0% B, from 6 to 14 min increasing from 0 to 40% in 8 min, then 40% for 5 min. Flow rate: 2 mL/min.

Separation were carried out on an Ascentis Express C30, 150 mm × 4.6 mm × 2.7 μm $_{d.p}$. Merck Life Science (Merck KGaA, Darmstadt, Germany. The used eluents were: A, CO_2; B CH_3OH; make-up solvent, CH_3OH; 35 °C was the column oven temperature and 150 bar was the regulator back pressure. The injection volume for standards was 3 μL. The MS was set as follows: Acquisition mode: SCAN in negative mode (−) and selected ion monitoring (SIM) (−). Interface temperature: 350 °C; DL temperature: 200 °C; block heater temperature: 200 °C; nebulizing gas flow (N_2) 3 L/min; drying gas flow (N_2) 5 L/min; Full scan range: 200–1200 m/z; event time: 0.05 sec for each event. The available standards, full scan, SIM, and multiple reaction monitoring (MRM) experiments were used for the apocarotenoid identifications. Transitions in the MS/MS experiments were previously optimized for the β-apocarotenals and apozeaxanthinals by Giuffrida et al. [20] and for ε-apoluteinals by Zoccali et al. [21]. β-Carotene and β-apo-8′-carotenal were quantitatively determined by multiple extractions as reported in Zoccali et al. [17]. Six-point calibration curves were constructed in the 0.1–20 mg L^{-1} range. The derived calibration curves had a coefficient of determination (R^2) of 0.9996 and 0.9991, respectively, for β-carotene and β-apo-8′-carotenal. Linearity was further confirmed using Mandel's fitting test. Limits of detection (LoD) were 0.03 and 0.04 mg L^{-1}, while limits of quantification (LoQ) were 0.091, 0.134 mg L^{-1}, respectively, for β-carotene and β -apo-8′-carotenal. Further, they were calculated by multiplying the standard deviation of the standard area at the lowest concentration level, three and ten times, respectively, and then were divided by the slope of the calibration curve.

3. Results and Discussion

Microalgae represents one of the most promising sources of bioactive molecules, including carotenoids [24,25]. In fact, they have the ability to adapt and grow in many different environmental conditions, going from tropic to temperate and artic waters [26]. In addition, many algae strains representing most habitats have stress handling mechanisms that frequently involve increased carotenoid production when exposed to unfavorable environmental conditions [27,28]. The actual knowledge of the carotenoids biosynthetic trails on microalgae is still mainly coming from plant studies [25].

The carotenoids composition of the selected four different microalgae species belonging to different botanical families and having different geographical origin—*Chlamydomonas* sp., *T. chuii*, *N. gaditana*, and *C. sorokiniana*—were reported in [29–34], although the selected psychrophilic *Chlamydomonas* sp. strain has not been previously explored. Interestingly, the possible occurrence of apocarotenoids in those microalgae species had never been investigated before. Extremophile species—in this case, the psychrophilic one—have mechanisms for tolerating conditions that would quickly kill other strains and probably have secondary metabolites not present in temperate species [35]. Some *Chlamydomonas* spp. and strains of *C. sorokiniana* have been reported to produce lutein as the main carotenoid [31,33]. *T. chuii* is a food approved species and has been reported to accumulate α and β-carotenes, whereas *N. gaditana*, which is frequently used in aquaculture feed due to its high eicosapentaenoic fatty acid (EPA) content, has been reported to accumulate violaxanthin and zeaxanthin [30,34].

The here reported methodology allowed for the determination of the native apocarotenoids prolife in four different microalgae species for the first time; a total of 29 different apocarotenoids, including various apocarotenoid fatty acid esters, were detected. The overall extraction and detection time for all the apocarotenoids was less than 10 min, including apocarotenoids esters, with an overall analysis time less than 20 min.

Table 1 shows the overall apocarotenoids detected by SFE-SFC-APCI(+/−)/QqQ MS analysis in the four microalgae strains. SIM detections and MRM transitions were applied to all the detected apocarotenoids except for the apo-violaxanthinals and apo-fucoxanthinals that were identified only using SIM detections, due to the lack of the respective standards.

Table 2 shows the overall apocarotenoids occurrence in the four microalgae strains. In general, it can be observed that the apocarotenoids were occurring in the microalgae strains in a scattered order although apo-12'-zeaxanthinal, β-apo-12'-carotenal, apo-12-luteinal, and apo-12'-violaxanthal were detected in all the investigated strains together with the two apocarotenoid esters, apo-10'-zeaxanthinal-C4:0, and apo-8'-zeaxanthinal-C8:0. The *Chlamydomonas* sp. strain showed the highest apocarotenoids occurrence among the investigated strains. In fact, 25 apocarotenoids were detected in this microalga. As far as we know this is the first detailed study on the apocarotenoids occurrence in any microalgae species. The presence of β-apo-8'-carotenal, β-apo-10'-carotenal, and apo-12'-violaxanthal were only previously reported by Sommella et al. [36] in Spirulina supplements. In Figure 3 are shown as example, the MRM analysis enlargements (transitions in APCI positive) relative to the detected β-apo-carotenals, apo-zeaxanthinals, and ε-apo-luteinals in the different microalgae strains. Further, it can be appreciated that all the different apocarotenoids were identified in less than 6 min of SFE-SFC-MS analysis. Although the purpose of this investigation was a qualitative apocarotenoids that profiled the four different microalgae strains, the available standards allowed us to also carry out a quantitative evaluation of the β-carotene and β-apo-8'-carotenal contents in the investigated samples. The amount of β-carotene was 89.7, 46.9, 20.6, and 4.2 ng mg^{-1} respectively in the *C. sorokiniana*, *N. gaditana*, *T. chuii*, and *Chlamydomonas* sp. samples, while β-apo-8'-carotenal was detected only in *C. sorokiniana* and *T. chuii* samples, with an amount of 0.7 and 0.5 ng mg^{-1}, respectively. Therefore, interestingly, considering the reported preliminary quantitative data the β-apo-8'-carotenal content was around the 0.8% and the 2.4% of the parent carotenoid in *C. sorokiniana* and *T. chuii*, respectively.

Table 1. Selected ion monitoring (SIM) *m/z*, Multiple reaction monitoring (MRM) with quantifier (Q) and qualifier (q) transitions (Collision Energy V) and Q/q % ratio of the detected apocarotenoids in the four microalgae strains.

Apocarotenoids	SIM (−)	MRM Transition (CE)		Q/q %
	m/z	Quantifier	Qualifier	
β-Apo-8′-Carotenal	416	+ 417>119 (−25)	+ 417>105 (−35)	73
β-Apo-10′-Carotenal	376	+ 377>105 (−35)	+ 377>119 (−30)	79
β-Apo-12′-Carotenal	350	+ 351>105 (−35)	+ 351>119 (−25)	74
β-Apo-14′-Carotenal	310	+ 311>105 (−25)	+ 311>119 (−25)	77
Apo-8′-Zeaxanthinal	432	+ 433>119 (−30)	+ 433>105 (−35)	95
Apo-10′-Zeaxanthinal	392	+ 393>105 (−35)	+ 393>119 (−25)	92
Apo-12′-Zeaxanthinal	366	+ 367>105 (−35)	+ 367>119 (−30)	80
Apo-14′-Zeaxanthinal	326	+ 327>105 (−35)	+ 327>119 (−30)	61
Apo-15′-Zeaxanthinal	300	+ 301>173 (−15)	+ 301>105 (−30)	57
Apo-8-Luteinal	432	+ 415>119 (−40)	+ 415>91 (−50)	95
Apo-10-Luteinal	392	+ 375>105 (−40)	+ 375>91 (−50)	91
Apo-12-Luteinal	366	+ 349>105 (−40)	+ 349>91 (−50)	90
Apo-14-Luteinal	326	+ 309>91 (−50)	+ 309>105 (−40)	55
Apo-15-Luteinal	300	+ 283>105 (−40)	+ 283>91 (−50)	95
Apo-8′-violaxanthin	448	n.d.	n.d.	
Apo-12′-violaxanthal	382	n.d.	n.d.	
Apo-14′-violaxanthal	342	n.d.	n.d.	
Apo-15′-violaxanthal	316	n.d.	n.d.	
Apo-8′-Fucoxanthinal	464	n.d.	n.d.	
Apo-10′-Fucoxanthinal	424	n.d.	n.d.	
Apo-14′-Fucoxanthinal	358	n.d.	n.d.	
Apo-15′-Fucoxanthinal	332	n.d.	n.d.	
Apocarotenoids-Esters	**SIM (−)**	**MRM transition (CE)**		
Apo-10′-Zeaxanthinal-C4:0	462	+ 463>105 (−40)	+ 463>119 (−35)	71
Apo-10′-Zeaxanthinal-C10:0	546	+ 547>105 (−35)	+ 547>119 (−30)	87
Apo-10′-Zeaxanthinal-C12:0	574	+ 575>105 (−35)	+ 575>119 (−30)	75
Apo-10′-Zeaxanthinal-C14:0	602	+ 603>105 (−40)	+ 603>119 (−30)	77
Apo-8′-Zeaxanthinal-C8:0	558	+ 559>105 (−40)	+ 559>119 (−40)	70
Apo-8′-Zeaxanthinal-C10:0	586	+ 587>119 (−40)	+ 587>105 (−40)	81
Apo-8′-Zeaxanthinal-C12:0	614	+ 615>105 (−40)	+ 615>119 (−40)	79

n.d. = not determined.

Table 2. Overall apocarotenoids occurrence in four microalgae strains.

Compound	Chlorella sorokiana NIVA-CHL 176	Nanochloropsis gaditana CCMP 526	Tetraselmis chui SAG 8-6	Chlamydomonas sp. CCMP 2294
Apo-8′-Zeaxanthinal	-	×	-	×
Apo-10′-Zeaxanthinal	×	-	-	×
Apo-12′-Zeaxanthinal	×	×	×	×
Apo-14′-Zeaxanthinal	×	-	×	×
Apo-15′-Zeaxanthinal	-	×	×	×
β-Apo-8′-Carotenal	×	-	×	-
β-Apo-10′-Carotenal	×	×	-	×
β-Apo-12′-Carotenal	×	×	×	×
β-Apo-14′-Carotenal	×	-	×	×
Apo-10′-Zeaxanthinal -C4:0	×	×	×	×
Apo-10′-Zeaxanthinal -C10:0	×	×	-	×
Apo-10′-Zeaxanthinal -C12:0	×	-	-	×
Apo-10′-Zeaxanthinal -C14:0	×	-	×	×
Apo-8′-Zeaxanthinal-C8:0	×	×	×	×
Apo-8′-Zeaxanthinal-C10:0	×	×	-	×
Apo-8′-Zeaxanthinal-C12:0	×	×	-	×
Apo-8-Luteinal	-	×	×	-
Apo-10-Luteinal	×	×	×	×
Apo-12-Luteinal	×	×	×	×
Apo-14-Luteinal	×	-	×	×
Apo-15-Luteina	×	-	-	×
Apo-8′-Violaxanthin	×	-	×	×
Apo-12′-Violaxanthal	×	×	×	×
Apo-14′-Violaxanthal	×	-	×	×
Apo-15′-Violaxanthal	-	×	×	×
Apo-8′-Fucoxanthinal	-	×	×	×
Apo-10′-Fucoxanthinal	-	×	-	-
Apo-14′-Fucoxanthinal	×	-	-	-
Apo-15′-Fucoxanthinal	×	-	×	×

× = Detected; -= not detected.

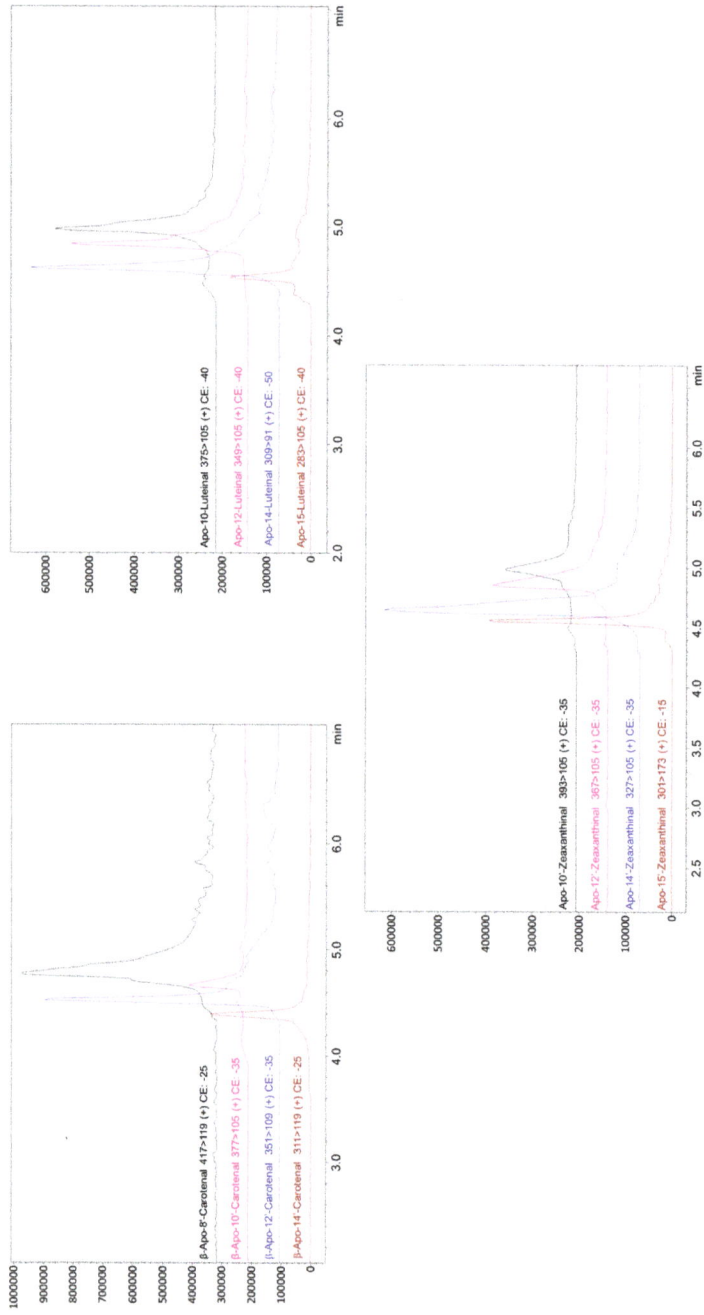

Figure 3. MRM analysis enlargements (transitions in APCI positive) relative to the detected β-apo-carotenals, apo-zeaxanthinals, and ε-apo-luteinals in the different microalgae strains.

4. Conclusions

The SFE-SFC-MS methodology applied in this work provided the first detailed report on the apocarotenoids detection and occurrence in four microalgae strains. The applied methodology was selective and efficient for the apocarotenoids detection. The reported determination of apocarotenoids in the microalgae further demonstrates the natural occurrence of these metabolites in the natural matrices, which certainly deserve further investigation. Moreover, the detection of fatty acids esterified apocarotenoids further demonstrate the wide occurrence and importance of the esterification process in carotenoids and carotenoid derivatives [37]. The possible exploitation of microalgae also containing biologically active apocarotenoids as functional food ingredients should be further explored by the food and feed industry.

Author Contributions: Conceptualization, D.G.; Formal analysis, M.Z.; Funding acquisition, P.D. and L.M.; Investigation, C.S.; Methodology, F.S.; Resources, K.S.; Writing-original draft, D.G.

Funding: We acknowledge the financial support from the NordForsk NCoE program, project "NordAqua" (project # 82845), and ERA-NET LAC program, project "SCREAM" (project # ELAC2014/BES0171).

Acknowledgments: Shimadzu and Merck KGaA Corporations.

Conflicts of Interest: No conflicts of interests by the authors.

References

1. Gong, M.; Bassi, A. Carotenoids from microalgae: A review of recent developments. *Biotechnol. Adv.* **2016**, *34*, 1396–1412. [CrossRef] [PubMed]
2. Matos, J.; Cardoso, C.; Bandarra, N.M.; Afonso, C. Microalgae as healthy ingredients for functional food: A review. *Food Funct.* **2017**, *8*, 2672–2685. [CrossRef] [PubMed]
3. Sathasivam, R.; Ki, J.-S. A review of the biological activities of microalgal carotenoids and their potential use in healthcare and cosmetic industries. *Mar. Drugs* **2018**, *16*, 26. [CrossRef] [PubMed]
4. Mc Gee, D.; Archer, L.; Paskuliakova, A.; Mc Coy, G.R.; Fleming, G.T.A.; Gillespie, E.; Touzet, N. Rapid chemotaxonomic profiling for the identification of high-value carotenoids in microalgae. *J. Appl. Phycol.* **2018**, *30*, 385–399. [CrossRef]
5. Maroneze, M.M.; Jacob-Lopez, E.; Queiroz Zepka, L.; Roca, M.; Perez-Galvez, A. Esterified carotenoids as new food components in cyanobacteria. *Food Chem.* **2019**, *287*, 295–302. [CrossRef] [PubMed]
6. Minhas, A.K.; Hodgson, P.; Barrow, C.; Adholeya, A. A review on the assessment of stress conditions for simultaneous production of microalgal lipids and carotenoids. *Front. Microbiol.* **2016**, *7*, 546. [CrossRef]
7. Liu, J.; Sun, Z.; Gerken, H. *Recent Advances in Microalgae Biotechnology*; OMICS Group eBooks: Foster City, CA, USA, 2014.
8. Britton, G.; Liaaen-Jensen, S.; Pfander, H. *Carotenoids Volume 5: Nutrition and Health*; Birkhauser: Basel, Switzerland, 2009.
9. Young, A.J.; Lowe, G.L. Carotenoids-Antioxidant Properties. *Antioxidants* **2018**, *7*, 28. [CrossRef]
10. Ceron-Garcia, M.C.; Gonzalez-Lopez, C.V.; Camacho-Rodriguez, J.; Lopez-Rosales, L.; Garcia-Camacho, F.; Molina-Grima, E. Maximizing carotenoid extraction from microalgae used as food additives and determined by liquid chromatography (HPLC). *Food Chem.* **2018**, *257*, 316–324. [CrossRef]
11. Abrahamsson, V.; Rodriguez-Meizoso, I.; Turner, C. Determination of carotenoids in microalgae using supercritical fluid extraction and chromatography. *J. Chromatogr. A* **2012**, *1250*, 63–68. [CrossRef]
12. Pour Hosseini, S.R.; Tavakoli, O.; Sarrafzadeh, M.H. Experimental optimization of SC-CO$_2$ extraction of carotenoids from Dunaliella salina. *J. Supercrit. Fluids* **2017**, *121*, 89–95. [CrossRef]
13. Hou, X.; Rivers, J.; Leon, P.; McQuinn, R.P.; Pogson, B.J. Synthesis and function of apocarotenoid signals in plants. *Trends Plant Sci.* **2016**, *21*, 792–803. [CrossRef] [PubMed]
14. Giuffrida, D.; Donato, P.; Dugo, P.; Mondello, L. Recent analytical techniques advances in the carotenoids and their derivatives determination in various matrixes. *J. Agric. Food Chem.* **2018**, *66*, 3302–3307. [CrossRef] [PubMed]
15. Eroglu, A.; Harrison, E.H. Carotenoid metabolism in mammals, including man: Formation, occurrence, and function of apocarotenoids. *J. Lipid Res.* **2013**, *54*, 1719–1730. [CrossRef] [PubMed]

16. Harrison, E.H.; Quadro, L. Apocarotenoids: Emerging roles in mammals. *Annu. Rev. Nutr.* **2018**, *38*, 153–172. [CrossRef] [PubMed]

17. Zoccali, M.; Giuffrida, D.; Dugo, P.; Mondello, L. Direct online extraction and determination by supercritical fluid extraction with chromatography and mass spectrometry of targeted carotenoids from Habanero peppers (*Capsicum chinense* Jacq.). *J. Sep. Sci.* **2017**, *40*, 3905–3913. [CrossRef] [PubMed]

18. Giuffrida, D.; Zoccali, M.; Arigò, A.; Cacciola, F.; Osorio-Roa, C.; Dugo, P.; Mondello, L. Comparison of different analytical techniques for the analysis of carotenoids in tamarillo (*Solanum betaceum* Cav.). *Arch. Biochem. Biophys.* **2018**, *646*, 161–167. [CrossRef] [PubMed]

19. Rodriguez, E.B.; Rodriguez-Amaya, D.B. Formation of apocarotenals and epoxycarotenoids from β-carotene by chemical reactions and by autoxidation in model system and processed foods. *Food Chem.* **2007**, *101*, 563–572. [CrossRef]

20. Giuffrida, D.; Zoccali, M.; Giofrè, S.V.; Dugo, P.; Mondello, L. Apocarotenoids determination in *Capsicum chinense* Jacq. cv Habanero, by supercritical fluid chromatography-mass spectrometry. *Food Chem.* **2017**, *231*, 316–323. [CrossRef]

21. Zoccali, M.; Giuffrida, D.; Salafia, F.; Giofrè, S.V.; Mondello, L. Carotenoids and apocarotenoids determination in intact human blood samples by online supercritical fluid extraction-supercritical fluid chromatography-tandem mass spectrometry. *Anal. Chim. Acta* **2018**, *1032*, 40–47. [CrossRef]

22. Guillard, R.R.L.; Hargraves, P.E. Stichochrysis immobilis is a diatom, not a chrysophyte. *Phycologia* **1993**, *32*, 234–236. [CrossRef]

23. Harris, E.H. *The Chlamydomonas Sourcebook. A comprehensive Guide to Biology and Laboratory Use*, 2nd ed.; Academic Press Inc.: San Diego, CA, USA, 1989.

24. Singh, S.; Kate, B.N.; Banerjee, U.C. Bioactive compounds from cyanobacteria and microalgae: An overview. *Crit. Rev. Biotechnol.* **2005**, *25*, 73–95. [CrossRef] [PubMed]

25. Galasso, C.; Corinaldesi, C.; Sansone, C. Carotenoids from marine organisms: Biological functions and industrial applications. *Antioxidants* **2017**, *6*, 96. [CrossRef] [PubMed]

26. Khan, M.I.; Shin, J.H.; Kim, J.D. The promising future of microalgae: Current status, challenges, and optimization of a sustainable and renewable industry for biofuels, feed, and other products. *Microb. Cell Fact.* **2018**, *17*, 36–46. [CrossRef] [PubMed]

27. Ahmed, F.; Fanning, K.; Netzel, M.; Turner, W.; Li, Y.; Schenk, P.M. Profiling of carotenoids and antioxidant capacity of microalgae from subtropical coastal and brackish waters. *Food Chem.* **2014**, *165*, 300–306. [CrossRef] [PubMed]

28. Skjånes, K.; Rebours, C.; Lindblad, P. Potential for green microalgae to produce hydrogen, pharmaceuticals and other high value products in a combined process. *Crit. Rev. Biotechnol.* **2013**, *33*, 172–215. [CrossRef] [PubMed]

29. Forján, E.; Garbayo, I.; Casal, C.; Vílchez, C. Enhancement of carotenoid production in *Nannochloropsis* by phosphate and sulphur limitation. *Commun. Curr. Res. Educ. Top. Trends Appl. Microbiol.* **2007**, *1*, 356–364.

30. Safafar, H.; van Wagenen, J.; Møller, P.; Jacobsen, C. Carotenoids, Phenolic Compounds and Tocopherols Contribute to the Antioxidative Properties of Some Microalgae Species Grown on Industrial Wastewater. *Mar. Drugs* **2015**, *13*, 7339–7356. [CrossRef] [PubMed]

31. Di Lena, G.; Casini, I.; Lucarini, M.; Lombardi-Boccia, G. Carotenoid profiling of five microalgae species from large-scale production. *Food Res. Int.* **2019**, *120*, 810–818. [CrossRef] [PubMed]

32. Montero, L.; Sedghi, M.; Garcia, Y.; Almeida, C.; Safi, C.; Engelen-Smit, N.; Cifuentes, A. Pressurized liquid extraction of pigments from *Chlamydomonas* sp. and chemical characterization by HPLC-MS/MS. *J. Anal. Test.* **2018**, *2*, 149–157. [CrossRef]

33. Cordero, B.F.; Obraztsova, L.; Couso, I.; Leon, R.; Vargas, M.A.; Rodriguez, H. Enhancement of lutein production in *Chlorella sorokiniana* (Chlorophyta) by improvement of culture conditions and random mutagenesis. *Mar. Drugs* **2011**, *9*, 1607–1624. [CrossRef] [PubMed]

34. Lubian, L.M. *Nannochloropsis* (Eustigmatophyceae) as source of commercially valuable pigments. *J. Appl. Phycol.* **2000**, *12*, 249–255. [CrossRef]

35. D'amico, S.; Collins, T.; Marx, J.C.; Feller, G.; Gerday, C. Psychrophilic microorganism: Challenges for life. *EMBO Rep.* **2006**, *7*, 385–389. [CrossRef] [PubMed]

36. Sommella, E.; Conte, G.M.; Salviati, E.; Pepe, G.; Bertamino, A.; Ostacolo, C.; Sansone, F.; del Prete, F.; Aquino, R.P.; Campiglia, P. Fast profiling of natural pigments in different Spirulina (Arthrospira platensis) dietary supplements by DI-FT-ICR and evaluation of their antioxidant potential by pre-column DPPH-UHPLC assay. *Molecules* **2018**, *23*, 1152. [CrossRef] [PubMed]

37. Hornero-Mendez, D. Occurrence of carotenoid esters in food. In *Carotenoid Esters in Food: Physical, Chemical and Biological Properties*; Mercadante, A.Z., Ed.; Royal Society of Chemistry: Cambridge, UK, 2019; Chapter 7; pp. 182–284.

antioxidants

MDPI

Review

Antioxidant Protection from UV- and Light-Stress Related to Carotenoid Structures

Gerhard Sandmann

Institute of Molecular Biosciences, Goethe-University Frankfurt/M, Max von Laue Str. 9, D-60438 Frankfurt, Germany; sandmann@bio.uni-frankfurt.de

Received: 19 June 2019; Accepted: 9 July 2019; Published: 11 July 2019

Abstract: This review summarizes studies of protection against singlet oxygen and radical damage by carotenoids. The main focus is on how substitutions of the carotenoid molecules determine high antioxidant activities such as singlet oxygen quenching and radical scavenging. Applied assays were carried out either in vitro in solvents or with liposomes, and in a few cases with living organisms. In the latter, protection by carotenoids especially of photosynthesis against light- and UV-stress is of major importance, but also heterotrophic organisms suffer from high light and UV exposure which can be alleviated by carotenoids. Carotenoids to be compared include C_{30}, C_{40} and C_{50} molecules either acyclic, monocyclic or bicyclic with different substitutions including sugar and fatty acid moieties. Although some studies are difficult to compare, there is a tendency towards mono and bicyclic carotenoids with keto groups at C-4/C-4' and the longest possible polyene structure functions to act best in singlet oxygen quenching and radical scavenging. Size of the carotenoid and lipophilic substituents such as fatty acids seem to be of minor importance for their activity but hydroxyl groups at an acyclic end and especially glycosylation of these hydroxyl groups enhance carotenoid activity.

Keywords: anti-oxidants; carotenoids; lipid peroxidation; light-stress; radicals; singlet oxygen; structure activity relationship; UV-stress

1. Oxidants and Antioxidants

Since the accumulation of oxygen in the atmosphere around 2 billion years ago caused by oxygenic photosynthesis, all living organisms have to cope with oxidative stress. Especially the reactive oxygen species (ROS), such as singlet oxygen $^1O_2$, hydroxyl radicals HO•, superoxide anion radicals $O_2^{•-}$ and hydrogen peroxide H_2O_2, generated by photosensitization or in cellular reactions exert their destructive power on the metabolism. Targeted metabolites are amino acids in enzymes, nucleic acids and fatty acids in lipid membranes [1]. $^1O_2$ is a very strong oxidant. It represents the electronically excited state of molecular oxygen with no unpaired electrons and is formed by energy transfer to the ground state from an excited photosensitizer. Photosensitizers are cell specific compounds such as porphyrins, riboflavin, chlorophylls or UV absorbing molecules [2]. The most efficient part of the solar spectrum is blue light and UV-B and UV-C radiation.

In the course of evolution, organisms have developed several strategies to protect from ROS [3]. They include enzymatic reactions to split H_2O_2 and antioxidants for hydrophilic (e.g., ascorbic acid) or lipophilic regions within the cell to inactivate HO• or quench $^1O_2$. The most prominent lipophilic antioxidants are tocopherols (vitamin E) and carotenoids. The latter pigments are also able to absorb the radiation energy from the photosensitizer preventing the transfer of excitation energy to ground state oxygen (Figure 1A). This is one way to prevent accumulation of $^1O_2$. Since the triplet energy level of carotenoids is close or below that of $^1O_2$, they can also efficiently drain the excitation energy from $^1O_2$. This absorption of excitation energy and its dissipation as heat is the principal mechanism of carotenoids to protect organisms from photosensitized formation and accumulation of $^1O_2$ [2,4].

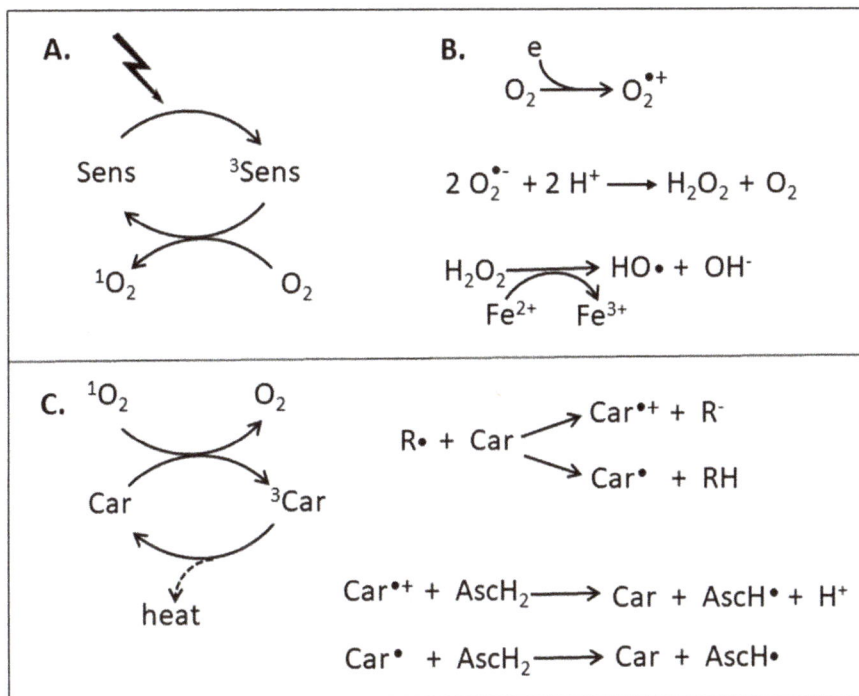

Figure 1. Formation of reactive oxygen species and their reaction with carotenoids. (**A**) Photosensitized formation of singlet oxygen, (**B**) radical formation, (**C**) reactions of carotenoids with singlet oxygen or radicals and regeneration by ascorbate.

A prominent function of carotenoids is the protection of the photosynthesis apparatus from damage under high light conditions [5]. This includes quenching of photosensitized triplet chlorophyll as first line of defense and of single oxygen once formed by reaction with triplet chlorophyll (Figure 1B). Another consequence of excess light is the formation of superoxide anion radicals $O_2\bullet^-$ by reduction of O_2 in the photosynthetic electron transport chain. Disproportionation of $O_2\bullet^-$ is a source of H_2O_2 (Figure 1B) [6]. Radicals such as $O_2\bullet^-$ and others including hydroxyl radical $HO\bullet$ which are formed from H_2O_2 by a Fenton-type reaction with Fe^{2+} or Cu^+ [7] can be inactivated by carotenoids. Several mechanisms are discussed by which carotenoids interact with radicals (Figure 1C). One type of reaction is adduct formation $R\text{-}Car\bullet$ by radical addition to the carotenoid polyene chain [8], another involves removal of an electron from the conjugated system of the carotenoid by the radical leading to the formation of a carotenoids radical cation $Car\bullet^+$ with a delocalized unpaired electron [2,9]. A third possibility is hydrogen abstraction from especially from allylic positions of the carotenoid yielding the radical $Car\bullet$ [10]. Finally, the carotenoid can be regenerated from a resonance-stabilized carotenoid radical by reaction with different reductants which is best studied for ascorbate [11,12]. In addition to radical scavenging by carotenoids as protective reaction, radical-carotenoid adducts are able to form peroxyl radicals under high oxygen pressure. Subsequently, the peroxyl radical acts as a generator of radicals in radical chain reactions involving for example oxidation of fatty acids in lipids [13,14]. However, this pro-oxidant reaction of carotenoids is minor under physiological oxygen concentrations. Not only in photoautotrophic but also in heterotrophic organisms, carotenoids are synthesized as antioxidants to protect especially against UV-generated ROS.

2. In Vitro Antioxidant Assays for Different Carotenoids

Several hundreds of carotenoid structures are known; they differ in their carbon chain lengths, the conjugated double bond system and the presence or absence of ionone rings with different substitutions [15]. Over the decades, their properties as radical scavengers and $^1O_2$ quenchers in relation to their structures were investigated. Experiments to determine their function in artificial systems were facilitated by the availability of novel carotenoids isolated from microorganisms or by their combinatorial synthesis through genetic engineering [16]. Furthermore, genetic modification of the carotenoid composition within organisms allowed the study of their functionality directly in living cells.

The applied assay systems for $^1O_2$ quenching or radical scavenging use different photosensitizers or radical generators, respectively, whereas $^1O_2$ or radical concentrations were measured directly or by oxidation of different substrates. This makes it difficult to compare the results and to rank the tested carotenoids [2]. Application of the carotenoids in the assays may be in fixed concentrations resulting in relative protection or by determination of the concentration values responsible for 50% quenching compared to a control without added carotenoid (IC_{50}). The assays can also be focused on $^1O_2$ formation versus $^1O_2$ peroxidation or radical formation versus peroxidation by radicals [17].

2.1. O₂ Quenching by Carotenoids

Most data on $^1O_2$ quenching were obtained with toluidine blue or methylene blue with similar chromophores as photosensitizers, and linoleic acid as the substrate to monitor $^1O_2$ formation. In this type of system, 18 cyclic and acyclic C_{40}-carotenoid structures differing in the number of conjugated double bonds and oxy group substitutions were investigated [18]. The authors show that the strongest protection effect is exerted with increasing numbers of conjugated double bonds. A smaller effect could be attributed to HO-groups and a major one to keto groups which extend the conjugated system as in astaxanthin. In a similar assay, several authors determined IC_{50} values for carotenoids to suppress $^1O_2$ accumulation. A comparison of the values for astaxanthin and its fatty acid esters gave no conclusive results. Either differences were very small [16] or the results varied with the use of the solvents [19]. In contrast to 100% ethanol, the IC_{50} was much higher in 50% hexane than in ethanol. The reason for the lower effectiveness of astaxanthin may be its poor solubility in the latter solvent mixture. Comparable IC_{50} values were also determined for a set of C_{30} 4,4'-diapocarotenoids [20–23] in comparison with the poorly protective C_{40} β-carotene and highly protective astaxanthin (Figure 2). The data show that C_{30} carotenoids although shorter are also efficient $^1O_2$ quenchers. Their antioxidant properties also gradually increase with the extension of the polyene chain from 10 via 11 to 13 conjugated double bonds including both keto groups in 4,4'-diapolycopendioc acid derivatives. Formation of its monoester decreased the quenching activity which was lowest with the diester. When comparing monocyclic carotenoids either with a β-ionone or a φ-ionone (= aromatic) ring, the aromatic end group enhanced $^1O_2$ quenching, whereas a C_{16} fatty acid bound to the sugar decreased this activity (Figure 2B) [24].

In an extensive study with chemically synthesized carotenoid analogues, their protective function on accumulation of chemically generated $^1O_2$ was investigated [25]. Among the natural carotenoids, rhodoxanthin with the same C_{40} carbon backbone and similar substitutions as canthaxanthin, but with 3,3'-diketo groups and 14 instead of 13 conjugated double bonds in retro configuration, exhibited the highest protection, followed by canthaxanthin, astaxanthin, lycopene and β-carotene (Figure 3A). It is notable that although naturally existing astaxanthin is a much better $^1O_2$ quencher than β-carotene, the synthetic C_{50} hydrocarbon decapreno β-carotene is more protective than the C_{50} astaxanthin analog. Results completely different to the studies mentioned above were obtained by Di Mascio et al. [26] with chemically generated $^1O_2$ and its quenching by carotenoids measured directly by infrared emission of $^1O_2$. Under these test conditions, lycopene, although with a shorter polyene system, was superior to astaxanthin and also γ-carotene was as effective as astaxanthin. Also, lutein, an α-carotene derivative with two hydroxyl groups, exhibited lower quenching than the unsubstituted hydrocarbon.

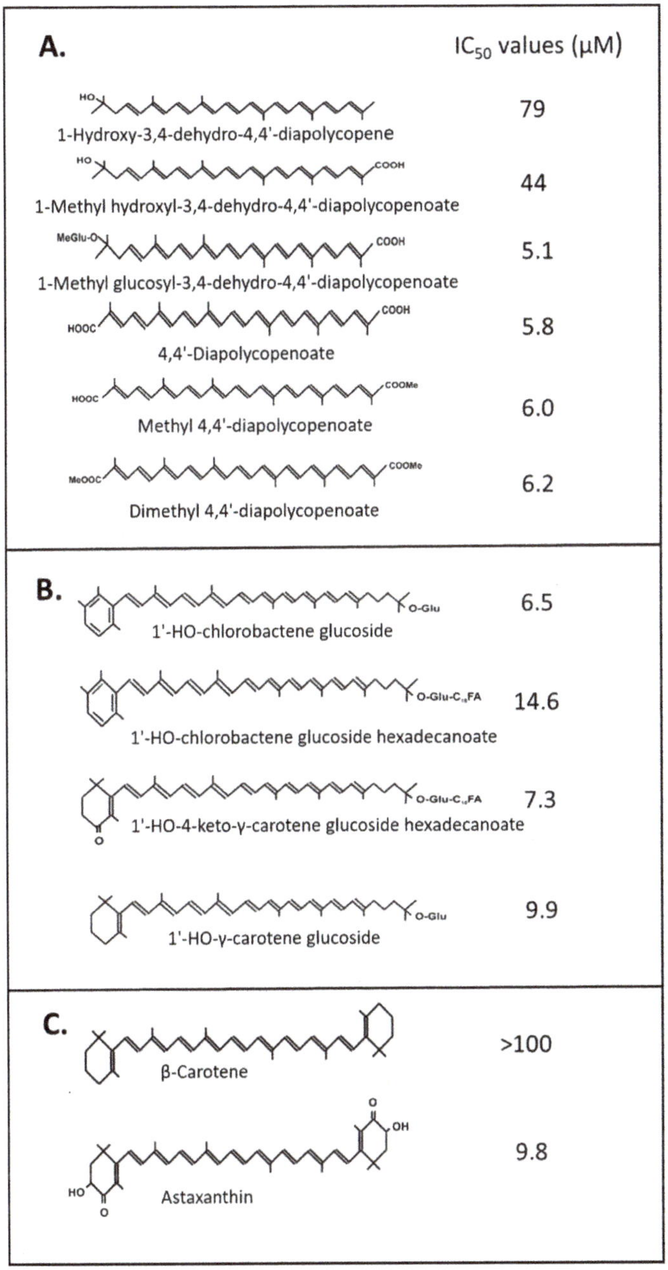

Figure 2. IC$_{50}$ values of carotenoids as $^1O_2$ quenchers assayed according to [20]. (**A**) C$_{30}$ 4,4′-diapolycopene derivatives [20–23], (**B**) monocyclic carotenoids [24], (**C**) astaxanthin and β-carotene for comparison [23].

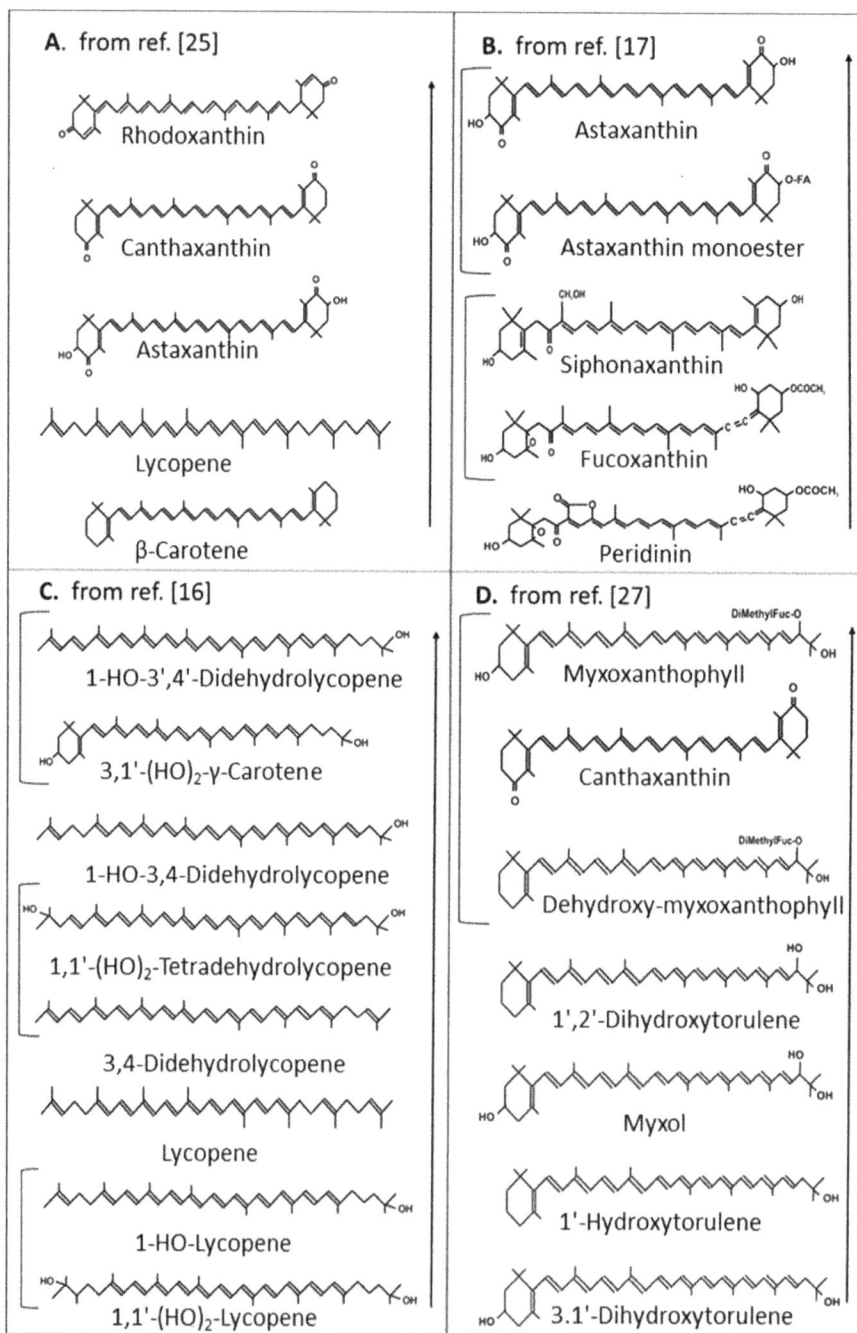

Figure 3. Comparison of $^1O_2$ quenching activities of different carotenoids. Arrows indicate increasing activity, brackets indicate similar activity.

Apart from assays determining protection by carotenoids on initial or early events of photosensitized peroxidation, other approaches focused on photosensitized reactions with artificial membranes. For a series of keto carotenoids, a comparison of both methods was made [17] showing that both provide equivalent results. Compared to astaxanthin with 13 conjugated double bonds with best protection, siphonaxanthin with 9 conjugated double bonds was less protective and fucoxanthin and peridinin with 8 plus one additional allenic double bond quenched even less (Figure 3B). Differences in their activity must be due to the different modification of the central carbon chain. Another study with the artificial membrane system compared lycopene with its hydroxyl derivatives and further desaturated polyene chains [16]. Among the carotenoids with the best protection against $^1O_2$ was 1-HO-3′,4′-didehydrolycopene with 13 conjugated double bonds and one hydroxyl group at C-1 followed by 1-HO-3,4-didehydrolycopene with 12 conjugated double bonds and the same hydroxyl group (Figure 3C). 3,4-Didehydrolycopene and 1,1′-(HO)$_2$-tetradehydrolycopene with 13 conjugated double bonds but differing from 1-HO-3′,4′-didehydrolycopene by the lack of the hydroxyl group or the presence of an additional one at C-1′ exhibited an even lower activity. Low activity in protection against $^1O_2$ was observed with unsubstituted lycopene which even gradually decreased with the presence of one or two hydroxyl groups at C-1 and C-1′. The only monocyclic carotenoid in this series was 3,1′-(HO)$_2$-γ-carotene with 11 conjugated double bonds and hydroxyl groups at C-3 of the β-ionone ring and at C-1 which had the same $^1O_2$ quenching potential as 1-HO-3′,4′-didehydrolycopene with 13 conjugated double bonds.

In photosynthetic organisms, chlorophyll is the dominant photosensitizer. Therefore, protection against chlorophyll-sensitized membrane oxidation was investigated with a set of monocyclic carotenoids together with canthaxanthin and lycopene for comparison [27]. Best protection was by canthaxanthin and both glycosides myxoxanthophyll and dehydroxy-myxoxanthophyll differing only by a 3-hydroxyl group (Figure 3D). Next in line with lower activity was 1′,2′-dihydroxytorulene the aglycone of dehydroxy-myxoxanthophyll followed by myxol the aglycone of myxoxanthophyll together with 1′-hydroxytorulene resembling 1′,2′-dihydroxytorulene without the 2′-hydroxyl group and 3,1′-dihydroxytorulene resembling the structure of 1′-hydroxytorulene without the 3-hydroxyl group. Lycopene was not protective in this series.

2.2. Radical Scavenging by Carotenoids

Radical scavenging by carotenoids was assayed with a radical generator either directly without a substrate (radical formation assay) or with the addition of phosphatidylcholine (PC) liposomes for lipid peroxidation (radical peroxidation assay). Radical generators used were either water soluble compounds such as 2,2′-azo-bis(2-amidinopropane) hydrochloride (AAPH) and 2,2′-azino-bis(3-ethylbenzothiazoline-6-sulphonic acid (ABTS) or lipid-soluble ones such as 2,2′-azo*bis*(2,4′-dimethylvaleronitrile) (AMVN), 2,2-azo-*bis*-isobutyronitrile (AIBN) and di(phenyl)-(2,4,6-trinitrophenyl) imino azanium (DPPH).

One investigation compares the scavenging activities in the radical formation assay versus the radical peroxidation assay with a set of algal keto carotenoids [17]. Especially fucoxanthin and peridinin both with an allenic double bond were much more efficient in protecting from peroxidation with ABTS cation radical than preventing lipid peroxidation initiated by DPPH, a stable free radical, whereas astaxanthin, its fatty acid monoester and siphonaxanthin inhibited well in both assays.

Radical peroxidation assays in the presence of phosphatidyl choline liposomes were carried out by several authors with different carotenes and hydroxylated and ketolated β-carotene derivatives. This included protection against radical initiation with AAPH or AMVN. Lipid-soluble AMVN generates peroxyl radicals by thermolysis which resembles the biological situation best. A study on scavenging with AMVN by carotenoids [28] showed that astaxanthin was superior to all other carotenoids tested followed by zeaxanthin, then canthaxanthin and finally β-carotene (Figure 4A). Under similar conditions but at a higher rate of peroxyl radical formation caused by a higher temperature, the order of carotenoids was not consistent [29] since here zeaxanthin and then β-cryptoxanthin were most efficient followed by β-carotene whereas the keto derivatives, echinenone, astaxanthin and canthaxanthin protected less

(Figure 4A). However, a high radical trapping by zeaxanthin and canthaxanthin and poor activity of β-carotene corresponds well to carotenoid oxidation with AAPH [30] or with HO• generation in a Fenton reaction by H_2O_2 and reduced iron [10]. Independent of the type of radical generation, carotenoid radical scavenging in lipid peroxidation is synergistic with other antioxidants such as α-tocopherol [29].

A.

with AMVN [28]

        Astaxanthin > Zeaxanthin > Canthaxanthin > β-Carotene

with AMVN [29]

        Zeaxanthin > β-Cryptoxanthin > β-Carotene > Echinenone > Astaxanthin >

        Canthaxanthin > Lycopene

with AAPH [25] or AMVN [30]

        Astaxanthin = Canthaxanthin > β-Carotene

B.

Neurosporene > β-Carotene > Zeaxanthin; Lycopene and ζ-Carotene without protection

C.

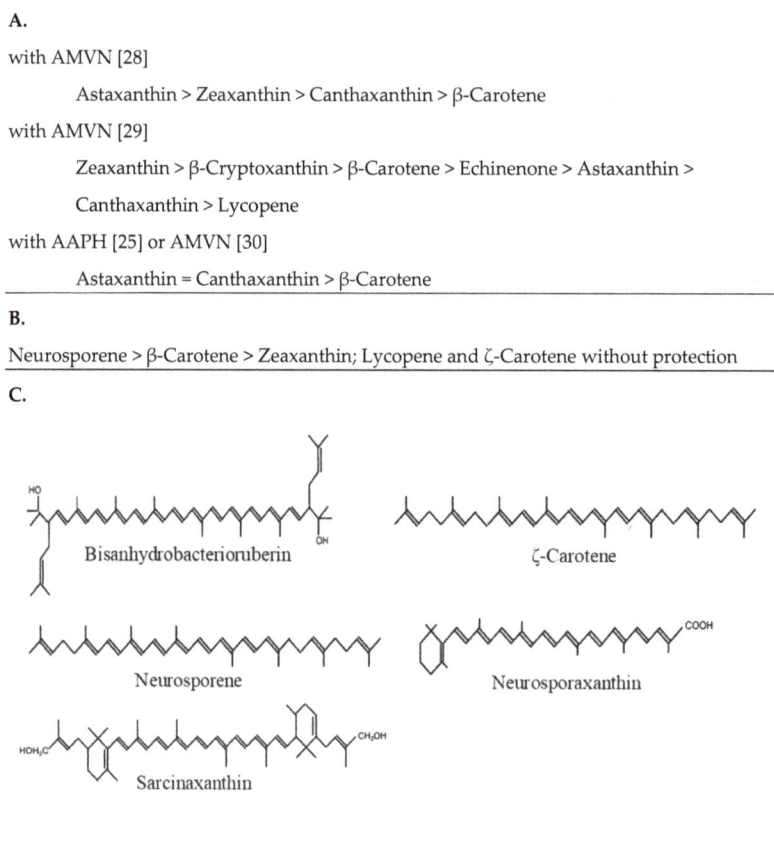

Figure 4. Protection by carotenoids against peroxyl radicals (**A**), against UV with α-terthienyl as the sensitizer (**B**), carotenoid structures mentioned in Section 3.1 and not shown in other figures (**C**). AMVN: 2,2′-*azobis*(2,4′-dimethylvaleronitrile); AAHP: 2,2′-azo-bis(2-amidinopropane) hydrochloride.

The same lycopene derivatives tested for $^1O_2$ quenching were also assayed for protection of radical lipid peroxidation started by AIBN [16]. The order of radical scavenging was the same as for the inactivation of $^1O_2$ with the exception of 1,1′-$(HO)_2$-tetradehydrolycopene which showed the lowest scavenging activity apart from 1,1′-$(HO)_2$-lycopene.

For the C_{50} bacterioruberin IC_{50} values for quenching of DPPH radical signal demonstrated that this carotenoid is 3-fold better in radical inactivation than β-carotene [31].

3. Protection in Living Organisms by Carotenoids against High Light and UV Radiation

Protection especially of the photosynthetic apparatus against photo oxidation and the formation of ROS is essential for autotrophic organisms. Nevertheless, also non-photosynthetic organisms are susceptible to high light and UV radiation and use carotenoids for protection [32]. Antioxidant activities of carotenoids can be tested in living organisms by quantitative or qualitative variation of their carotenoid composition. In animal studies, carotenoids such as β-carotene, canthaxanthin or astaxanthin injected or fed with the diet exerted antioxidant activity (see [33] for an overview). In microorganisms, testing of carotenoids can be better achieved by use of pigment mutants, inhibitors of carotenoid biosynthesis and genetic pathway modification followed by enhancement of oxidative stress either by application of a photosensitizer or by an oxidant.

3.1. Heterotrophic Organisms

One of the first studies to demonstrate carotenoid protection against photodynamic damage was carried out with *Micrococcus luteus* (formerly *Sarcina lutea*), a bacterium containing the C_{50} carotenoid sarcinaxanthin [34] as well as with mutants with lower or zero carotenoid content, by direct sunlight exposure [35] or after adding a photosensitizer [36]. In each case, the number of survival cells of the non-pigmented mutant was orders of magnitude lower than the pigmented wild type. Similar results were obtained with *Curtobacterium flaccumfaciens* (formerly *Corynebacterium poinsettiae*) containing C_{50} bisanhydrobacterioruberin [37] after inhibition of carotenogenesis and with its non-pigmented mutant. This approach also demonstrated that the action of the photosensitizer was oxygen dependent [38]. C_{30} 4,4′-diapolycopene-dioic acid ester also protects growth of a *Bacillus* species against peroxide-induced peroxidation [39].

Carotenoids are known to protect against UV-B radiation in fungi, but they can also play a role as substrates for formation of cleavage products with functions in fungal metabolism. Treatment of strains of *Ustilago violacea* with visible and UV radiation showed that carotenogenic strains containing neurosporene and lycopene [40] were more resistant to cell death caused by light, demonstrating the effectiveness of carotenoids as protectants [41]. Similar studies carried out with the neurosporaxanthin-synthesizing fungus *Neurospora crassa* gave comparable results [42]. The importance of carotenoids in photoprotection is supported by the light-dependent up-regulation of the synthesis of these pigment in many fungal species [43].

Non-carotenogenic *Escherichia coli* is susceptible to near UV radiation [44]. Transformants carrying a plasmid for the synthesis of zeaxanthin diglycoside had a much higher survival rate. The protection by this carotenoid was much stronger when UV treatment was carried out in the presence of the UV-sensitizer α-terthienyl [45]. When *E. coli* was engineered to synthesize different carotenoids in a similar concentration range, the carotenes neurosporene followed by β-carotene were the most protective against UV treatment but also zeaxanthin showed some degree of protection unlike lycopene and ζ-carotene, especially when oxidative pressure was increased with α-terthienyl (Figure 4B) [46]. Zeaxanthin acted concentration dependent and zeaxanthin diglycoside was less effective even in higher concentrations.

Oxidative stress in humans and health benefits of carotenoids were reviewed recently [47]. Extended exposure of light causes skin damage in animals. Photosensitivity of the skin is strongly pronounced in light-sensitive porphyria in which porphyrin is accumulated as an endogenous photosensitizer triggering the generation of $^{1}O_2$. In an experiment with albino mice, the application of hematoporphyrin and exposure to black light with short UV radiation phases was lethal to most individuals [48]. Injection of β-carotene prior to treatment resulted in a significantly higher number of surviving animals. Patients with erythropoietic protoporphyria suffer from a disturbed porphyrin metabolism causing an accumulation of protoporphyrin in the body. This makes the skin extra susceptible to photosensitation. Indigestion of a α- and β-carotene diet over several weeks resulted in an improved tolerance for light with a suppression of the burning symptoms [49]. These results demonstrate that in vivo carotenoids can alleviate skin damage caused by porphyrin photosensitation by quenching either its exited state or the resulting $^{1}O_2$. However, a high dose of β-carotene tested on isolated keratinocytes from the skin led to pro-oxidant action [50]. This pro-oxidant effect demonstrated in an animal model with high β-carotene application in combination with smoke exposure may also be responsible for increased risk for smokers of lung cancer when treated with high carotenoid doses [51].

3.2. Photosynthetic Organisms

In diatoms, fucoxanthin is the major carotenoid of the chloroplast. By inhibition of its synthesis to about half the concentration of the control, the alga became highly susceptible to high light and peroxide treatment [17]. In photosynthetic algae and plants, carotenoids in the chloroplast not only function as antioxidants but are also constituents of pigment protein complexes functioning in photosynthetic electron transport and especially in light-harvesting [52]. In plant photosynthesis, carotenoids are not only protective but also act as accessory pigments for light-harvesting. This light-harvesting function is replaced in cyanobacteria by phycocyanin [53]. Therefore, for investigations focusing on carotenoid antioxidant action in oxygenic photosynthesis, cyanobacteria are the most useful organisms since modification of carotenoid composition avoids disturbance of light-harvesting protein complexes and their function. Consequently, carotenoid action can be mostly attributed to protection of photosynthesis or chlorophyll levels during the influence of light or UV radiation.

The unicellular canobacterium *Synechocystis* synthesizes a combination of myxoxanthophyll, β-carotene, echinenone and zeaxanthin. By inactivation of two genes of the carotenoid pathway, deletion mutants have been generated which lack either echinenone formation, zeaxanthin formation or the synthesis of both carotenoids, simultaneously. With these mutants, the effects of high light inhibition of photosynthetic activity and methylene blue sensitized chlorophyll oxidation were determined [54]. Lack of echinenone had only a moderate effect on both parameters, but lack of zeaxanthin strongly decreased photosynthesis and chlorophyll content. When both carotenoids were missing, *Synechocystis* suffered considerably more. *Synechococcus* is another unicellular cyanobacterium which only accumulates β-carotene and zeaxanthin. By transformation, its zeaxanthin content was either increased or converted to canthaxanthin [55,56]. The influence of the modified carotenoid composition on photosynthesis under high light or treatment with near UV is shown in Table 1. High zeaxanthin in the cells prevented a decrease of photosynthesis in high light as well as under UV. Prevention of UV damage was directly correlated to the zeaxanthin content of the transformants. A replacement of zeaxanthin by canthaxanthin almost completely protected photosynthesis in high light. From the investigations with these cyanobacteria, it can be concluded that of the carotenoids present in cyanobacteria, canthaxanthin is the best protectant of photosynthesis against high light and UV radiation followed by zeaxanthin and then by echinenone. The significance of these carotenoids in the protection of photosynthesis in cyanobacteria is supported by the light-dependent up-regulation of canthaxanthin synthesis [57] or zeaxanthin synthesis [58] in different species. In plant photosynthesis, the major protecting carotenoid is zeaxanthin which is generated by de-epoxidation of violaxanthin in the xanthophyll cycle under high light conditions. Due to the UV defense attributed to zeaxanthin, tobacco plants were genetically engineered to synthesize zeaxanthin in larger quantities [59]. Upon treatment with UV or $^1O_2$ generated by the dye rose bengal as a photosensitizer, a decrease in photosynthetic oxygen evolution was lower and lipid peroxidation less pronounced than in the wild-type with lower zeaxanthin content. With tobacco as a higher plant model, it could be demonstrated that enhanced zeaxanthin protects photosynthesis under radiation stress leading to a higher biomass.

Table 1. Protection of photosynthetic oxygen evolution in *Synechoccocus* transformants with increased zeaxanthin or canthaxanthin content after exposure to high light or UV-B [55,56].

Transformants	Carotenoid (mg/g dw)	Photosynthesis (% of Untreated Control)	
		High Light *	UV Treatment **
non-transgenic	Zeax 1.0	66	32
CrtB	Zeax 1.1	–	37
Psy	Zeax 1.7	–	58
CrtZ	Zeax 1.9	82	60
Bkt ***	Canth 1.6+ Zeax 0.7	95	–

* 6 h exposure to high light (900 µE m^{-2} s^{-1}); ** 6 h exposure to UV-B (6.8 W/m^{-2}); Canth: canthaxanthin; Zeax: zeaxanthin, *** Sandmann unpublished.

4. Conclusion on Carotenoids as Antioxidants

The antioxidant function of carotenoids as quenchers of excited photosensitizers and $^1O_2$, as well as their radical scavenging potential, is well-documented. The degree of these activities depends on their chemical structures. The in vitro assays to assess their antioxidant potential can be divided into those targeting $^1O_2$ formation and peroxidation by $^1O_2$ or radical formation and peroxidation by radicals. Among the carotenoids for which IC_{50} values for $^1O_2$ inactivation are available, substituted acyclic C_{30} carotenoid diacids showed the highest activities (Figure 2A). In general, superior carotenoids for $^1O_2$ quenching are those with the longest polyene system corresponding to the lowest triplet energy levels which is especially the case for rhodoxanthin (Figure 3A) and for HO-chlorobactene glucoside with an aromatic end group (Figure 2B). Substituents affect the electronic density in the polyene chain. There is no clear tendency of electron withdrawing hydroxyl groups at acyclic carotenoids (Figure 3C). They negatively affect $^1O_2$ quenching activity of acyclic carotenoids (Figure 3C) but seem to be favorable when located at α-position to the polyene chain of monocyclic carotenoids (Figure 3D). Hydroxyl groups at C-3 of the β-ionone ring somehow decrease quenching activity. However, glycosylation of a hydroxyl group strongly enhanced $^1O_2$ quenching (Figure 3D). When the sugar moiety is esterified with a fatty acid, activity decreased again (Figure 2A,B). In a similar way, esterification of the carboxylic groups of the C_{30} carotenoids lowers the quenching activity (Figure 2A). The comparison of dihydroxy-γ-carotene to dihydroxylycopene both with 13 conjugated double bonds indicated that a monocyclic structure is more favorable for $^1O_2$ quenching than an acyclic structure (Figure 3C). An important substituent at the β-ionone rings for high protection against $^1O_2$ are keto groups at positions C-4 and C-4′ which extend the conjugated double bond system (Figure 3A,B). Only a very few studies demonstrating $^1O_2$ quenching by carotenoids in non-photosynthetic organisms are available. Transgenic *E. coli* synthesized varying carotenoids. The $^1O_2$ quenching activity of these transformants differed fundamentally from the in vitro results. In this bacterium, neurosporene with only 9 double bonds protected best and better than lycopene, and even zeaxanthin, against photosensitized UV treatment (Figure 4B). However, in cyanobacteria with oxygenic photosynthesis, protection of different carotenoids against light and UV damage in general resembled the carotenoid activities of the in vitro results (Table 1).

Substantial radical scavenging in vitro was obtained with most of the carotenoids which also protect against $^1O_2$. One way for natural carotenoids to react with radicals is by hydrogen abstraction from the allylic carbon such as C-3 in canthaxanthin and C-2 in rhodoxantin or from the 3-hydroxyl group of astaxanthin. This ensures the best resonance stabilization of the resulting carotenoid radical due to the extended polyene system allowing for delocalization over most of the carotenoid molecule. For zeaxanthin with the allylic carbon at C-4, the polyene system is shorter and radical delocalization less pronounced corresponding to a somehow lower radical inactivation activity. In the case of electron capture by a radical from the carotenoid (Figure 1C), the polyene chain is as important for the resonance stabilization of the resulting carotenoid radical cation as for the carotenoid radical generated by H• abstraction.

Either for $^1O_2$ quenching or radical inactivation, the length of the carotenoid seems of minor importance since among C_{30}, C_{40} and C_{50} derivatives highly potent species can be found. Polar sugar moieties or carboxylic groups at the end of the carotenoid molecule enhance quenching activity (Figure 3B) whereas fatty acid substituents at the sugar reverse this effect (Figure 2A,B). This points at an advantageous anchoring with these polar groups to the hydrophilic outside of the membrane [60] but with a free floating tail in the hydrophobic core of the membrane. However, the best tested $^1O_2$ quenching carotenoid is the synthetic C_{50} hydrocarbon decapreno β-carotene without any polar groups [25] which should be completely mobile in the membrane. Therefore, positioning and orientation of carotenoids in the lipid membrane seems to be less crucial for their antioxidant function than generally discussed [29].

The in vitro studies provide detailed information on the structural properties of carotenoids which are responsible for maximum protection against $^1O_2$ formation and inactivation and for making them good radical quenchers. Only a few investigations on antioxidant action of carotenoids in bacteria and fungi are available. They demonstrate that carotenoids protect against radiation and oxidative stress but are not sufficient to recognize optimized structures. In photo autotrophic organisms, damage on photosynthesis is mainly caused by excited chlorophyll leading to formation of $^1O_2$. By genetic modification of the carotenoid composition it was evident that diketo β-carotene (canthaxanthin) is superior to dihydroxy β-carotene (zeaxanthin) and that β-carotene is of minor importance (Table 1). These results correspond well to those obtained in vitro with chlorophyll as a photosensitizer (Figure 3D).

Funding: This research received no external funding.

Conflicts of Interest: The author declares no conflict of interest.

References

1. Santos, A.L.; Oliveira, V.; Baptista, I.; Henriques, I.; Gomes, N.C.; Almeida, A.; Correia, A.; Cunha, A. Wavelength dependence of biological damage induced by UV radiation on bacteria. *Arch. Microbiol.* **2013**, *195*, 63–74. [CrossRef]
2. Edge, R.; McGarvey, D.J.; Truscott, T.G. The carotenoids as antioxidants—A review. *J. Photochem. Photobiol. B Biol.* **1997**, *41*, 189–200. [CrossRef]
3. Sies, H. Strategies of antioxidant defense. *Eur. J. Biochem.* **1993**, *215*, 213–219. [CrossRef] [PubMed]
4. Ogilby, P.R. Singlet oxygen: There is still something new under the sun, and it is better than ever. *Photochem. Photobiol. Sci.* **2010**, *9*, 1543–1560. [CrossRef] [PubMed]
5. Young, A.J.; Lowe, G.M. Antioxidant and prooxidant properties of carotenoids. *Arch. Biochem. Biophys.* **2001**, *385*, 20–27. [CrossRef] [PubMed]
6. Marklund, S. Spectrophotometric study of spontaneous disproportionation of superoxide anion radical and sensitive direct assay for superoxide dismutase. *J. Biol. Chem.* **1976**, *251*, 7504–7507.
7. Kehrer, J.P. The Haber–Weiss reaction and mechanisms of toxicity. *Toxicology* **2000**, *149*, 43–50. [CrossRef]
8. Mortensen, A.; Skibsted, L.H.; Willnow, A.; Everett, S.A. Kinetics of photobleaching of β-carotene in chloroform and formation of transient carotenoid species absorbing in the near infra-red. *Free Rad. Res.* **1998**, *28*, 69–80. [CrossRef]
9. Liebler, D.C.; McClure, T.D. Antioxidant reactions of β-carotene: Identification of carotenoid-radical adducts. *Chem. Res. Toxicol.* **1996**, *9*, 8–11. [CrossRef]
10. Woodall, A.A.; Lee, S.W.; Weesie, R.J.; Jackson, M.J.; Britton, G. Oxidation of carotenoids by free radicals: Relationship between structure and reactivity. *Biochim. Biophys. Acta* **1997**, *1336*, 33–42. [CrossRef]
11. Burke, M.; Edge, R.; Land, E.J.; McGarvey, D.J.; Truscott, T.G. One-electron reduction potentials of dietary carotenoid radical cations in aqueous micellar environments. *FEBS Lett.* **2001**, *500*, 132–136. [CrossRef]
12. Böhm, F.; Edge, R.; Truscott, T.G. Interactions of dietary carotenoids with activated (singlet) oxygen and free radicals: Potential effects for human health. *Mol. Nutr. Food Res.* **2012**, *56*, 205–216. [CrossRef] [PubMed]
13. Krinsky, N.I.; Yeum, K.J. Carotenoid-radical interactions. *Biochem. Biophys. Res. Commun.* **2003**, *305*, 754–760. [CrossRef]
14. Yeum, K.-J.; Aldini, G.; Russel, R.R.; Krinsky, N.I. Antioxidant/pro-oxidant actions of carotenoids. In *Carotenoids Nutrition and Health*; Birkhäuser Verlag: Basel, Switzerland, 2009; Volume 5, pp. 235–268.
15. Britton, G.; Liaan-Jensen, S.; Pfander, H. *Carotenoid Handbook*; Birkhäuser Verlag: Basel, Switzerland, 2004.
16. Albrecht, M.; Takaichi, S.; Steiger, S.; Wang, Z.-Y.; Sandmann, G. Novel hydroxycarotenoids with improved antioxidative properties produced by gene combination in Escherichia coli. *Nat. Biotechnol.* **2000**, *18*, 843–846. [CrossRef] [PubMed]
17. Dambeck, M.; Sandmann, G. Antioxidative activities of algal keto carotenoids acting as antioxidative protectants in the chloroplast. *Photochem. Photobiol.* **2014**, *90*, 814–819. [CrossRef]
18. Hirayama, O.; Nakamura, K.; Hamada, S.; Kobayashi, Y. Singlet oxygen quenching ability of naturally occurring carotenoids. *Lipids* **1994**, *29*, 149–150. [CrossRef]

19. Kobayashi, M.; Sakamoto, Y. Singlet oxygen quenching ability of astaxanthin esters from the green alga Haematococcus pluvialis. *Biotechnol. Lett.* **1999**, *21*, 265–269. [CrossRef]
20. Osawa, A.; Iki, K.; Sandmann, G.; Shindo, K. Isolation and identification of 4,4'-diapolycopene-4,4'-dioic acid produced by *Bacillus firmus* GB1 and its singlet oxygen quenching activity. *J. Oleo Sci.* **2013**, *62*, 955–960. [CrossRef]
21. Osawa, A.; Ishii, Y.; Sasamura, N.; Morita, M.; Köcher, S.; Müller, V.; Sandmann, G.; Shindo, K. Hydroxy-3,4-dehydro-apo-8'-lycopene and methyl hydroxy-3,4-dehydro-apo-8'-lycopenoate, novel C (30) carotenoids produced by a mutant of marine bacterium *Halobacillus halophilus*. *J. Antibiot.* **2010**, *63*, 291–295, reprinted in *J. Antibiot.* **2014**, *67*, 733–735. [CrossRef]
22. Shindo, K.; Endo, M.; Miyake, Y.; Wakasugi, K.; Morritt, D.; Bramley, P.M.; Fraser, P.D.; Kasai, H.; Misawa, N. Methyl glucosyl-3,4-dehydro-apo-8'-lycopenoate, a novel antioxidative glyco-C (30)-carotenoic acid produced by a marine bacterium *Planococcus maritimus*. *J. Antibiot.* **2008**, *61*, 729–735, reprinted in *J. Antibiot.* **2014**, *67*, 731–733. [CrossRef]
23. Shindo, K.; Asagi, E.; Sano, A.; Hotta, E.; Minemura, N.; Mikami, K.; Tamesada, E.; Misawa, N.; Maoka, T. Diapolycopenedioic acid xylosyl esters A, B, and C, novel antioxidative glyco-C$_{30}$-carotenoic acids produced by a new marine bacterium *Rubritalea squalenifaciens*. *J. Antibiot.* **2008**, *61*, 185–191. [CrossRef] [PubMed]
24. Osawa, A.; Kasahara, A.; Mastuoka, S.; Gassel, S.; Sandmann, G.; Shindo, K. Isolation of a novel carotenoid, OH-chlorobactene glucoside hexadecanoate, and related rare carotenoids from *Rhodococcus* sp. CIP and their antioxidative activities. *Biosci. Biotechnol. Biochem.* **2011**, *75*, 2142–2147. [CrossRef] [PubMed]
25. Beutner, S.; Bloedorn, B.; Frixel, S.; Hernandez Blanco, I.; Hoffmann, T.; Martin, H.-D.; Mayer, B.; Noack, P.; Ruck, C.; Schmidt, M.; et al. assessment of antioxidant properties of natural colorants and phytochemicals: Carotenoids, flavonoids, phenols and indigoids. The role of β-carotene in antioxidant functions. *J. Sci. Food Agric.* **2001**, *81*, 559–568. [CrossRef]
26. Di Mascio, P.; Kaiser, S.; Sies, H. Lycopene as the most efficient biological carotenoid singlet oxygen quencher. *Arch. Biochem. Biophys.* **1989**, *274*, 532–538. [CrossRef]
27. Gildenhoff, N. Isolation von Myxoxanthophyll und Seinen Derivaten und ihre Bedeutung beim Schutz vor Photooxidation. Master's Thesis, Goethe University, Frankfurt, Germany, 2005. (Unpublished results).
28. Lim, B.P.; Nagao, A.; Terao, J.; Tanaka, K.; Suzuki, T.; Takama, K. Antioxidant activity of xanthophylls on peroxyl radical-mediated phospholipid peroxidation. *Biochim. Biophys. Acta* **1992**, *1126*, 178–184. [PubMed]
29. Woodall, A.A.; Britton, G.; Jackson, M.J. Carotenoids and protection of phospholipids in solution or in liposomes against oxidation by peroxyl radicals: Relationship between carotenoid structure and protective ability. *Biochim. Biophys. Acta* **1997**, *1336*, 575–586. [CrossRef]
30. Palozza, P.; Krinsky, N.I. Astaxanthin and canthaxanthin are potent antioxidants in a membrane model. *Arch. Biochem. Biophys.* **1992**, *297*, 291–295. [CrossRef]
31. Zalazar, L.; Pagola, P.; Miro, M.V.; Churio, M.S.; Cerletti, M.; Martınez, C.; Iniesta-Cuerda, M.; Soler, A.J.; Cesari, A.; De Castro, R. Bacterioruberin extracts from a genetically modified hyperpigmented Haloferax volcanii strain: Antioxidant activity and bioactive properties on sperm cells. *J. Appl. Microbiol.* **2019**, *126*, 796–810. [CrossRef]
32. Krinsky, N.I. Non-photosynthetic functions of carotenoids. *Phil. Trans. R. Soc. Lond. B.* **1978**, *248*, 581–590. [CrossRef]
33. Palozza, P.; Krinsky, N.I. Antioxidant effect of carotenoids in vivo and in vitro: An overview. *Methods Enzymol.* **1992**, *213*, 403–420.
34. Liaaen-Jensen, S.; Weeks, O.B.; Strang, R.H.; Thirkell, D. Identity of the C-50-carotenoid dehydrogenans-P439 and sarcinaxanthin. *Nature* **1967**, *214*, 379–380. [CrossRef] [PubMed]
35. Mathews-Roth, M.M.; Krinsky, N.I. Sudies on the protective function of the carotenoid pigments of Sarcina lutea. *Photochem. Photobiol.* **1970**, *11*, 419–428. [CrossRef] [PubMed]
36. Mathews, M.M.; Sistrom, W.R. The function of carotenoid pigments of Sarcina lutea. *Arch. Microbiol.* **1960**, *35*, 139–146. [CrossRef]
37. Norgard, S.; Aasen, A.J.; Liaaen-Jensen, S. Bacterial carotenoids. XXXII. C$_{50}$-Carotenoids. 6. Carotenoids from *Corynebacterium poinsettiae* including for new C$_{50}$-diols. *Acta Chem. Scand.* **1970**, *24*, 2183–2197. [CrossRef] [PubMed]
38. Kunisawa, R.; Stanier, R.Y. Studies on the role of carotenoid pigments in a chemoheterotrophic bacterium Corynebacterium poinsettiae. *Arch. Microbiol.* **1958**, *31*, 146–156. [CrossRef]

39. Steiger, S.; Perez-Fons, L.; Fraser, P.D.; Sandmann, G. Biosynthesis of a novel C$_{30}$ carotenoid in Bacillus firmus isolates. *J. Appl. Microbiol.* **2011**, *13*, 888–895. [CrossRef] [PubMed]

40. Garber, E.D.; Baird, M.L.; Chapman, D.J. Genetics of *Ustilago violacea* I: Carotenoid mutants and carotenogenesis. *Botanical Gazette* **1975**, *136*, 341–346. [CrossRef]

41. Will III, O.H.; Jankowski, P.; Kowacs, A.; Rossing, W.; Schneider, P.; Newland, N.A. A comparison of photo-killing among carotene and cytochrome c accumulating strains of the smut fungus *Ustilaga violacea* at specific wavelengths from 400 to 600 nm. *Photochem. Photobiol.* **1987**, *45*, 609–615. [CrossRef]

42. Rau, W. Mechanism of photoregulation of carotenoid biosynthesis in plants. *Pure Appl. Chem.* **1985**, *57*, 777–784. [CrossRef]

43. Blanc, P.L.; Tuveson, R.W.; Sargent, M.L. Inactivation of carotenoid-producing and albino strains of *Neurospora crassa* by visible light, blacklight, and ultraviolet radiation. *J. Bacteriol.* **1976**, *125*, 616–625. [PubMed]

44. Tuveson, R.W.; Larson, R.A.; Kagan, J. Role of cloned carotenoid genes expressed in *Escherichia coli* in protecting against inactivation by near-UV light and specific phototoxic molecules. *J. Bacteriol.* **1988**, *170*, 4675–4680. [CrossRef] [PubMed]

45. Kagan, J.; Bazin, M.; Santus, R. Photosensitization with alpha-terthienyl: The formation of superoxide ion in aqueous media. *J. Photochem. Photobiol. B.* **1989**, *3*, 165–174. [CrossRef]

46. Sandmann, G.; Kuhn, S.; Böger, P. Evaluation of structurally different carotenoids in *Escherichia coli* transformants as protectants against UV-B radiation. *Appl. Environ. Microbiol.* **1998**, *64*, 1972–1974. [PubMed]

47. Bohn, T. Carotenoids and markers of oxidative stress in human observational studies and intervention trials: Implications for chronic diseases. *Antioxidants* **2019**, *8*, 179. [CrossRef] [PubMed]

48. Mathews, M.M. Protective effect of β-carotene against lethal photosensitization by hæmatoporphyrin. *Nature* **1964**, *203*, 1092. [CrossRef] [PubMed]

49. Mathews, M.M. Carotenoids in erythropoietic protoporphyria and other photosensitivity diseases. *Ann. N. Y. Acad. Sci.* **1993**, *691*, 127–138. [CrossRef] [PubMed]

50. Lotan, S.B.; Vitt, K.; Scholz, P.; Keck, C.M.; Meinke, M.C. ROS production and glutathione response in keratinocytes after application of β-carotene and VIS/NIR irradiation. *Chem. Biol. Interact.* **2018**, *280*, 1–7.

51. Russel, R.M. The enigma of β-carotene in carcinogenesis: What can be learned from animal studies? *J. Nutr.* **2004**, *134*, 262S–268S. [CrossRef] [PubMed]

52. Telfer, A.; Pascal, A.; Gall, A. Carotenoids in photosynthesis. In *Carotenoids Natural Functions*; Birkhäuser Verlag: Basel, Switzerland, 2008; Volume 4, pp. 266–306.

53. Glaser, A.N. Light harvesting by phycobilisomes. *Ann. Rev. Biophys. Biophys. Chem.* **1985**, *14*, 47–77. [CrossRef]

54. Schäfer, L.; Vioque, A.; Sandmann, G. Functional in situ evaluation of photosynthesis-protecting carotenoids in mutants of the cyanobacterium Synechocystis PCC6803. *J. Photochem. Photobiol. B Biol.* **2005**, *78*, 195–201. [CrossRef]

55. Götz, T.; Windhövel, U.; Böger, P.; Sandmann, G. Protection of Photosynthesis against Ultraviolet-B Radiation by Carotenoids in Transformants of the CyanobacteriumSynechococcus PCC7942. *Plant Physiol.* **1999**, *120*, 599–604.

56. Albrecht, M.; Steiger, S.; Sandmann, G. Expression of a ketolase gene mediates the synthesis of canthaxanthin in *Synechococcus* leading to tolerance against photoinhibition, pigment degradation and UV-B sensitivity of photosynthesis. *Photochem. Photobiol.* **2001**, *73*, 551–555. [CrossRef]

57. Schöpf, L.; Mautz, J.; Sandmann, G. Multiple ketolases involved in light regulation of canthaxanthin biosynthesis in Nostoc punctiforme PCC 73102. *Planta* **2013**, *237*, 1279–1285. [CrossRef] [PubMed]

58. Steiger, S.; Schäfer, L.; Sandmann, G. High-light upregulation of carotenoids and their antioxidative properties in the cyanobacterium *Synechocystis* PCC 6803. *J. Photochem. Photobiol. B Biol.* **1999**, *521*, 14–18. [CrossRef]

59. Götz, T.; Sandmann, G.; Römer, S. Expression of a bacterial carotene hydroxylase gene (*crtZ*) enhances UV tolerance in tobacco. *Plant Mol. Biol.* **2002**, *50*, 129–142. [CrossRef] [PubMed]

60. Havaux, M. Carotenoids as membrane stabilizers in chloroplasts. *Trend Plant Sci.* **1998**, *3*, 147–151. [CrossRef]

antioxidants

MDPI

Review

Nutritional Importance of Carotenoids and Their Effect on Liver Health: A Review

Laura Inés Elvira-Torales [1,2,*], Javier García-Alonso [1] and María Jesús Periago-Castón [1,*]

[1] Department of Food Technology, Food Science and Nutrition, Faculty of Veterinary Sciences, Regional Campus of International Excellence "Campus Mare Nostrum", Biomedical Research Institute of Murcia (IMIB-Arrixaca-UMU), University Clinical Hospital "Virgen de la Arrixaca", University of Murcia, Espinardo, 30071 Murcia, Spain

[2] Department of Food Engineering, Tierra Blanca Superior Technological Institute, Tierra Blanca 95180, Mexico

*   Correspondence: lauraines.elvira@um.es (L.I.E.-T.); mjperi@um.es (M.J.P.-C.)

Received: 30 June 2019; Accepted: 18 July 2019; Published: 19 July 2019

Abstract: The consumption of carotenoids has beneficial effects on health, reducing the risk of certain forms of cancer, cardiovascular diseases, and macular degeneration, among others. The mechanism of action of carotenoids has not been clearly identified; however, it has been associated with the antioxidant capacity of carotenoids, which acts against reactive oxygen species and inactivating free radicals, although it has also been shown that carotenoids modulate gene expression. Dietary carotenoids are absorbed and accumulated in the liver and other organs, where they exert their beneficial effects. In recent years, it has been described that the intake of carotenoids can significantly reduce the risk of suffering from liver diseases, such as non-alcoholic fatty liver disease (NAFLD). This disease is characterized by an imbalance in lipid metabolism producing the accumulation of fat in the hepatocyte, leading to lipoperoxidation, followed by oxidative stress and inflammation. In the first phases, the main treatment of NAFLD is to change the lifestyle, including dietary habits. In this sense, carotenoids have been shown to have a hepatoprotective effect due to their ability to reduce oxidative stress and regulate the lipid metabolism of hepatocytes by modulating certain genes. The objective of this review was to provide a description of the effects of dietary carotenoids from fruits and vegetables on liver health.

Keywords: β-carotene; lycopene; lutein; β-cryptoxanthin; non-alcoholic fatty liver disease (NAFLD); hepatic steatosis

---

1. Introduction

In recent decades, carotenoids (lycopene, β-carotene, lutein, zeaxanthin, and β-cryptoxanthin) have aroused great interest in the field of human nutrition, as they act as biological antioxidants, contributing to the defense of the organism against reactive oxygen species (ROS) [1,2] and play a protective role in conditions, such as diabetes and CVD [3], impacting cellular signaling pathways and influencing the expression of certain genes, and inhibiting specific enzymes involved in the development of certain forms of cancer [4]. Dietary carotenoids are mainly accumulated in the liver, where they are transferred to be transported by the different lipoproteins for their release into the blood circulation and thus to be deposited and stored in different organs and tissues, such as the kidneys, adipose tissue, adrenal glands, testes, skin, and the prostate [5]. Adipose tissue (abdominal fat) is an important reserve site for carotenoids, showing a strong association between the intake of carotenoids and the concentrations of these antioxidants in plasma [6]. The carotenoids present in the skin can protect against the damaging effects of radiation and neutralize the attacks of free radicals, particularly ROS [7,8]. In addition, the concentrations of these carotenoids in the skin can increase with their dietary supplementation and decrease in people with oxidative stress, such as smokers. Similarly, carotenoids in plasma and skin decrease with exposure to UV rays [9]. The accumulation of these antioxidants, as well as of their metabolites in the liver, can exert a positive effect on the hepatocyte metabolism, regulating the cellular oxidative state in certain liver pathologies. Non-alcoholic fatty liver disease (NAFLD) is currently considered one of the most frequent chronic liver diseases in the world and represents a serious and growing clinical problem in developed and developing countries [10,11]. NAFLD can occur in different states, from simple steatosis to non-alcoholic steatohepatitis (NASH) with liver fibrosis and cirrhosis, which can eventually lead to hepatocellular carcinoma [12–14]. Ingestion of carotenoid-type antioxidants through the diet is considered as one of the possible mechanisms in the treatment of non-alcoholic fatty liver disease (NAFLD), thus avoiding the progression of NASH and other types of liver diseases [13,15–19]. In this review, we focused on the preventive potential of carotenoids at a nutritional level and their effect on liver health.

2. Carotenoids

Carotenoids are lipophilic pigments synthesized by plants, fungi, algae, and bacteria [20,21]. In plants, carotenoids contribute to the photosynthetic system and protect them against photodamage, in addition to helping in the production of phytohormones [22]. As pigments, they are responsible for the red, orange, pink, and yellow colors of the leaves of plants, fruits, vegetables, and some birds, insects, fish, and crustaceans [23–26]. More than 750 types of carotenoids have been identified in nature, but only about 100 are present in detectable amounts within the human diet [20]. Between 30 and 40 carotenoids have been found in human blood samples and the six most abundant carotenoids consist of more than 95% of the carotenoids found in blood plasma: lycopene, lutein, β-carotene, β-cryptoxanthin, α-carotene, and zeaxanthin [27,28].

2.1. Chemical Structure

Typically, carotenoids are composed of forty carbon atoms formed by the union of eight isoprene units covalently linked. These structures can be completely linear or have rings at one or both ends, and these rings can contain hydroxyl groups, ketones, epoxies, or others. Carotenoids belong to two structural groups: carotenes containing carbon and hydrogen atoms and xanthophylls containing at least one oxygen atom [29,30]. In addition, carotenoids can be classified into two categories: carotenoids with provitamin A activity (β-carotene and β-cryptoxanthin) and carotenoids without provitamin A activity (lycopene and lutein) [31]. The different carotenoids originate basically by modification in the base structure, cyclization of the final groups, and by the introduction of oxygen groups that give them their characteristic colors and antioxidant properties [23].

2.2. Source of Carotenoids

Fruits and vegetables are the main sources of carotenoids in the human diet, providing 80–90% of these compounds in developed countries and 82% in developing countries [32,33]. Since carotenoids cannot be synthesized in the human body, they are used as biomarkers to reflect the intake of fruits and vegetables, establishing a direct relationship between the consumption of vegetables and the concentration of carotenoids in blood [34,35].

Carotenoids are found in almost all foods of plant origin, but Britton and Khachik [36] established a classification of dietary sources according to their carotenoid content, establishing sources with a low content (0–0.1 mg/100 g fresh product), moderate (0.1–0.5 mg/100 g fresh product), high (0.5–2 mg/100 g fresh product), and very high content (>2 mg/100 g fresh product).

β-carotene is the main carotenoid present in the human diet. It is found mainly in yellow–orange and dark green fruits and vegetables, such as carrots, squash, spinach, papaya, mango, apricots, and sweet potatoes [37,38]. Lycopene is a carotenoid that lacks provitamin A activity and is responsible for red to pink colors in fruits and vegetables, such as tomatoes, red grapefruit, watermelon, apricots, pink guava, and papaya [12,39]. Tomatoes and tomato-based products are the most common sources of lycopene in the human diet and account for more than 85% of the dietary intake of this carotenoid in

North America [13]. Likewise, in the European diet, the intake of tomato-based lycopene and tomato products (canned tomatoes, mashed potatoes, soups, and tomato sauces) constitutes 57% in France, 56% in the Republic of Ireland and the United Kingdom, 61% in the Netherlands, and 97% in Spain [40].

Lutein is a non-provitamin A carotenoid that belongs to the family of xanthophylls or oxycarotenoids [13]. It is distributed in a wide variety of vegetables, such as kale, spinach, and winter squash, and fruits such as mango, papaya, peaches, plums, and oranges [41]. Commercially, lutein is extracted from the flowers of the tagetes (*Tagetes erecta* L.), which contains 0.1–0.2% of carotenoids, of which 80 are diesters of lutein [42,43]. β-Cryptoxanthin, a xanthophyll with pro-vitamin A activity and one of the lesser-known carotenoids, is usually present in pumpkins, peppers, carrots, oranges, peaches, tangerines, and in tropical fruits such as papaya [39,44–46]. Table 1 shows the content of the six most important carotenoids in the diet for different fruits and vegetables [1,47,48].

Table 1. Data on the contents of major carotenoids in fruits and vegetables common in the human diet (mg/100 g) [1,47,48].

Food	Lutein	Zeaxanthin	β-Cryptoxanthin	α-Carotene	β-Carotene	Lycopene
Avocado	0.21–0.36	0.01	0.02–0.03	0.02–0.03	0.05–0.08	–
Banana	0.09–0.19	–	n.d.–0.01	0.06–0.16	0.04–0.13	n.d.–0.25
Peach	–	–	–	–	–	0.01
Guava	–	–	0.02–0.12	n.d.	0.10–2.67	0.77–1.82
Fig	0.08	–	0.01	0.02	0.04	0.32
Kiwi	–	–	–	–	<0.02	<0.01
Mandarin Orange	–	–	0.63–1.06	n.d.	0.11–0.32	–
Mango	–	–	0.02–0.32	n.d.	0.11–1.20	<0.01–0.72
Apple	0.02	n.d.	n.d.	n.d.	0.019	n.d.
Passion fruit	–	–	0.18	–	0.36–0.78	–
Orange	–	–	0.07–0.14	n.d.	0.17–0.48	n.d.
Peach	–	0.02–0.04	0.004–0.02	–	0.14–0.26	–
Papaya	0.09–0.32	–	n.d.–1.03	n.d.	0–08–0.66	n.d.–7.56
Pineapple	–	–	0.07–0.12	n.d.	0.14–0.35	0.27–0.61
Watermelon	–	–	n.d.	n.d.	0.31–0.78	4.77–13.52
Grapefruit	–	–	–	–	–	0.75
Tangerine	0.17	il	0.43	0.03	0.26	–
Grape	0.01	n.d.	n.d.	n.d.	0.02	n.d.
Plum	0.08–0.09	n.d.	n.d.	n.d.	0.09–0.14	n.d.
Apricot	0.12–0.19	n.d.–0.04	–	n.d.–0.04	0.59–3.80	0.05
Chard	3.60	0.01	n.d.	n.d.	2.90	n.d.
Artichoke	0.59–0.63	–	–	–	0.27–0.37	–
Broccoli	0.71–3.30	–	n.d.	n.d.	0.29–1.75	n.d.
Pumpkin	0.63	–	0.06	–	0.49	0.50
Sweet Potato	0.05	–	–	–	7.83	–
Peas	1.91	il	n.d.	n.d.	0.52	n.d.
Red Pepper	0.25–8.51	0.59–1.35	0.25–0.45	n.d.–0.29	1.44–2.39	–
Jalapeño Pepper	0.84	–	–	0.01–0.17	0.38–8.58	–
Spinach	5.93–7.90	il	n.d.	n.d.	3.10–4.81	n.d.
Lettuce	1.00–4.78	–	–	–	0.87–2.96	–
Corn	0.41	0.22	n.d.	n.d.	n.d.	n.d.
Cucumber	0.46–0.84	il	n.d.	n.d.	0.11–0.27	n.d.
Red chili	n.d.	–	–	–	6.53–15.40	–
Cabbage	0.45	il	n.d.	n.d.	0.41	n.d.
Tomato	0.05–0.21	il	n.d.	n.d.	0.32–1.50	0.85–12.70
Carrot	0.25–0.51	il	n.d.	2.84–4.96	4.35–8.84	n.d.
Kale	4.80–11.47	–	–	–	1.02–7.38	–
Parsley	6.40–10.65	il	n.d.	n.d.	4.44–4.68	n.d.
Coriander	6.00–14.80	–	–	2.90–11.30	4.80–8.40	–

–: not included in the references, n.d.: not detected or quantified, il: included in lutein.

The composition and content of carotenoids in fruits and vegetables is very variable, and depends on factors such as variety, genotype, season, geographical location/climatic conditions, soil, maturity stage, type of processing, and storage conditions [1,49,50]. Generally, carotenoid content in foods is not altered by common methods of cooking at home (microwave cooking, steaming, and boiling), but extreme heat can cause the oxidative destruction of carotenoids [51].

2.3. Bioavailability and Bioaccessibility

Before absorption, the carotenoids must be extracted from the food matrix in which they are ingested, transferred to the lipid emulsion, and incorporated into the micelles containing pancreatic lipases and bile salts [26,38,52]. Then, carotenoids are able to be transported to the enterocytes. However, their bioaccessibility in plants foods is remarkably low and these compounds are characterized by a slow absorption rate, since their chemical structure interacts deeply with the macromolecules within the food matrix of plants [53].

The factors that influence the bioavailability and bioaccessibility of carotenoids can be classified into two groups: (i) those related to carotenoids, which include dosage, chemical structure (isomeric forms), and interactions between carotenoids, and (ii) those not related to carotenoids, which include food processing and storage (raw, dehydrated, frozen, cooked), food composition, particle size of digested food, consumer biometrics, and transportation efficiency through the enterocyte [1,54–59]. Among these unrelated factors, thermal treatment increases the accessibility and bioavailability of carotenoids, due to the rupture of the cell walls and links with other macromolecules facilitating the release of carotenoids and improving their absorption [60–62]. The increase in the bioavailability of carotenoids when processing temperatures are above 100 °C (canning and sterilization) has been associated with isomerization, since the *cis* (Z) isomers are more bioavailable [1]. Several investigations have addressed the fact that the Z-isomerization of carotenoids influences not only bioavailability, but also antioxidant, anticancer, and antiatherosclerotic activities [63]. However, these results differ according to the type of carotenoid. For example, the Z-isomers of lycopene and astaxanthin have a greater bioavailability than the all-E-isomers [64,65], whereas the Z-isomers of β-carotene and lutein have a lower bioavailability than the all-E-isomers [66] It is important to understand the effect of E/Z-isomerization on functional changes, since this depends on its bioavailability and functionality of the carotenoids by ingestion [63]. In addition, mechanical processing, such as chopping and chewing, helps reduce the size of the particles and releases carotenoids from chloroplasts and tissues, increasing their bioavailability [67–69].

Another factor that influences the bioavailability of these compounds is the presence of other components of the diet. Thus, the presence of fat has a positive effect, and an intake of 3 to 5 g of fat is essential for the optimal absorption of carotenoids since it favors their incorporation into the micelle, facilitating their subsequent absorption [70,71]. Previous studies have even shown that long-chain fatty acids, such as oleic acid, are more beneficial for the absorption of non-polar carotenoids (carotenes) than non-polar ones (xanthophylls), by favoring their incorporation into the micelle [72,73]. On the other hand, the presence of dietary fiber and protein binding negatively affects their accessibility. Dietary fiber decreases the absorption of carotenoids by trapping them and interacting with bile acids, which leads to an increase in faecal excretion of fat and fat-soluble substances, such as carotenoids [57,74]. Protein–carotenoid complexes (such as lutein and zeaxanthin in spinach) and the microcrystalline form of some carotenoids (such as lycopene in tomatoes or β-carotene in carrots) makes them less available compared to those that are completely submerged in lipid droplets [75,76].

To assess bioavailability, the physiological state of the consumer must also be evaluated. The bioavailability of carotenoids can be modified in parasitic infestations (by intestinal helminths) and when there are diseases that produce intestinal dysfunction, as alterations in the uptake of carotenoids and in bioconversion have been observed [1]. In addition, age appears to be another factor contributing to the bioavailability of carotenoids, with a direct relationship between plasma carotenoid concentrations and the consumption of plant foods in groups of young adults, but not in advanced age groups [77], which suggests a lower bioavailability associated with age.

2.4. Nutritional Requirements

To maintain a high content of carotenoids in the diet, dietary sources, the factors that influence their bioavailability, and the frequency of intake must be considered. As discussed above, serum carotenoid concentrations are used as a biomarker to establish the dietary intake of fruits and vegetables [78–80]. However, dietary intake and serum carotenoid levels show high variability between subjects from different populations, as well as between individuals from the same population [81,82], which may be due to the geographical availability of fruits and vegetables, socioeconomic status, and cultural factors [83]. Thus, in European countries, the total intake of carotenoids varies from 9.5 to 16 mg/day (3 to 6 mg/day for β-carotene), vegetables and fruits being the main dietary sources [84]. In the United States, the average intake of lycopene varies from 6.6 to 10.5 mg/day for men and 5.7–10.4 mg/day for women, of which more than 85% of the intake comes from tomatoes and tomato products (salsa, pasta, soup, juice, and ketchup) [85]. The intake of carotenoids in diets in various countries is shown in Table 2 [40,84,86].

Table 2. Dietary consumption of carotenoids in different countries (data are reported as mean and [median]) [40,84,86].

Sample (N), Country	Woman/Man (Age)	Dietary Intake (mg/day)					
		α-car	β-car	β-cryp	Lut/ Zea	Lyco	Total
EUROPE							
N = 1968, Italy	W, M (> 1)	0.15	2.6	0.17	4.01	7.38	14.31
N=75, France	W, M (25–45)	[0.74]	[5.84]	[0.45]	[2.50]	[4.75]	14.28
N = 65, North Ireland	W, M (25–45)	1.04	5.55	0.99	1.59	5.01	14.18
N = 71, United Kingdom	W, M (25–45)	[1.04]	[5.55]	[0.99]	[1.59]	[5.01]	14.18
N = 73, Ireland	W, M (25–45)	1.23	5.16	0.78	1.56	4.43	13.16
N = 72, Netherlands	W, M (25–45)	0.68	4.35	0.97	2.01	4.86	12.87
N = 159, Sweden	W (56–75)	1.03	3.47	0.46	2.64	2.15	9.75
N = 3000, Spain	W, M (18–64)	0.27	1.46	0.32	1.24	3.06	6.35
OCEANIA							
N = 91, Australia	W (18–70)	[2.0]	[6.87]		[2.28]	[5.05]	16.2
AMERICA							
N = 459, Costa Rica	115 W (59±10)	0.73	4.67	0.55	2.89	5.77	14.61
	344 M (56±11)	0.45	3.41	0.38	2.41	5.45	12.10
N = 402, USA (Afro-American)	155 M (34–84)	[0.33]	[2.21]	[0.11]	[1.85)	[3.16]	7.66
	247 W (34–84)	[0.25]	[2.21]	[0.13]	[1.93]	[2.60]	7.12
N = 50, Dominican Republic	W, M (50–90)	0.7	2.7	0.22	1.33	1.46	6.41
USA	W, M (≥ 20)	0.4	1.9	0.2	1.4	1.4	5.3
N = 55,950, Brazil	W, M (≥ 10)	0.16	0.92	0.16	0.83	0.83	2.9

Some in vivo studies have shown that exposure to high doses of carotenoids has a pro-oxidant effect. The beta-carotene and retinol efficacy trial (CARET) showed that participants who received a combination of β-carotene (30 mg) and vitamin A (25,000 IU retinyl palmitate) had incidences of lung cancer and mortality. These results are like those found for β-carotene in the alpha-tocopherol, beta-carotene (ATBC) study performed on 29,133 male smokers in Finland [87,88]. Haider et al. [89] also observed that high concentrations of β-carotene (50 μM) in primary pneumocyte type II cells produced a cytotoxic effect.

No specific recommendations regarding the intake of carotenoids have been published, apart from a daily recommendation for provitamin A carotenoids in the case of not consuming other sources of this vitamin. Thus, healthy adults are advised to consume 10.8–21.6 mg/day in order to reach the recommended daily dose of retinol (900 to 700 µg of equivalents/day) [90]. For the other carotenoids, some recommendations have been published according to their beneficial effect on health. Grune et al. [91] proposed a daily consumption of 7 mg of β-carotene to cover the basic need for this carotenoid. For lycopene, an intake of 5 to 7 mg per day was recommended for healthy people to maintain the circulating levels of this carotenoid, in order to combat oxidative stress and prevent chronic diseases [92]. Heath et al. [93] reported that higher concentrations of lycopene (35–75 mg/day) may be required when there is a disease, such as cancer and cardiovascular diseases. A daily intake level of up to 10 mg of lutein and zeaxanthin was recommended for the treatment of age-related early macular degeneration [94]. An intake of 3 mg/day of β-cryptoxanthin was suggested for the treatment of patients with NAFLD [17]. However, these recommendations are based on intervention studies and were proposed by researchers. Official dietary recommendations have not yet been issued by public health organizations.

3. Carotenoids and Hepatic Health

3.1. Pathogenesis of Non-Alcoholic Fatty Liver Disease (NAFLD)

The liver is the largest viscera in the body (weighing about 1.5 kg in a healthy adult) and is involved in numerous metabolic processes, such as the regulation of carbohydrates, lipids, and proteins. It also performs specific functions, such as synthesis of steroid hormones, detoxification of drugs, and conjugation of bilirubin [95].

The most common diseases of the liver are due to viral infections, alcohol consumption, autoimmune diseases, ischemia, and genetic disorders [96]. Obesity is also associated with liver damage and with an increased risk of NAFLD [97]. This disease affects 25%–45% of the general population and has a higher prevalence in diabetic and obese patients. Recent research has shown that in the United States, more than a third of adults and 17% of young people are obese. Among these, 70%–80% have NAFLD [98]. The prevalence of NAFLD in South America (evaluated by ultrasound) was estimated at around 30.45% and seems to be higher than the rate reported for the United States (20.0–29.9%). A meta-analysis published in 2016 reported an average prevalence of 23.71% in Europe, varying from 5–44% in different countries [99]. It is estimated that in the next 20 years, NAFLD will become the main cause of morbidity and mortality related to the liver and will be one of the main causes for liver transplantation [100].

NAFLD refers to the accumulation of excess fat in more than 5% of hepatocytes, without significant alcohol intake [101], and ranges from steatosis, with inflammation, to the progression to non-alcoholic steatohepatitis (NASH), fibrosis, cirrhosis, and in some cases, hepatocellular carcinoma [102]. The underlying mechanism for the progression of steatosis to inflammation and fibrosis is not fully understood, although insulin resistance, lipid metabolism disorders and oxidative stress are implicated [103–105]. Currently, the hypothesis that was proposed to explain the pathogenesis of NASH defends the existence of two impacts or hits; the first hit is due to insulin resistance and lipid overload, which leads to simple hepatic steatosis, and the second hit involves oxidative stress, lipid peroxidation, the induction of proinflammatory cytokines, and the inflammation process, which are the main causes that lead to the presence of NASH [104,106,107]. Lipid overload is caused by an increased entry of free fatty acids (FFA), leading to de novo lipogenesis. Insulin resistance, associated with the metabolic syndrome, also increases the accumulation of liver fat by increasing the release of free fatty acids and simulating anabolic processes [104,108]. The excess of FFA is stored in droplets within the hepatocyte, resulting in steatosis, which induces the innate immune response, with the recruitment of immune cells such as macrophages and T cells. As a result of the excess of intracellular fat and the deterioration of mitochondrial oxidative capacity, oxidation occurs in peroxisomes and microsomes, causing an increase in lipid peroxidation that leads to the generation of reactive oxygen species (ROS), damaging the proteins and the DNA. Kupffer cells (liver macrophages) produce proinflammatory cytokines, such as tumor necrosis factor α (TNF-α), in response to oxidative stress, mediating the inflammatory response that can cause cell death and damage. Thus, oxidative stress, together with inflammation, leads to fibrogenesis, a fundamental feature of the progression of steatosis to NASH [109].

The mechanisms that represent the pathogenesis of NAFLD are presented in Figure 1 [19,110]. Although the pathogenesis and evolution of NAFLD to NASH is known, there is no agreement on the most effective pharmacological agents for its treatment. However, antioxidants such as carotenoids can play an important role in the defense against oxidative stress by avoiding or delaying oxidation, by neutralizing free radicals by sequestering singlet oxygen and inhibiting the progression of steatosis to steatohepatitis [2]. In fact, several studies have mentioned that carotenoids such as β-carotene, lycopene, lutein, and β-cryptoxanthin have antioxidant effects against lipid peroxidation in the liver of rats [111–113].

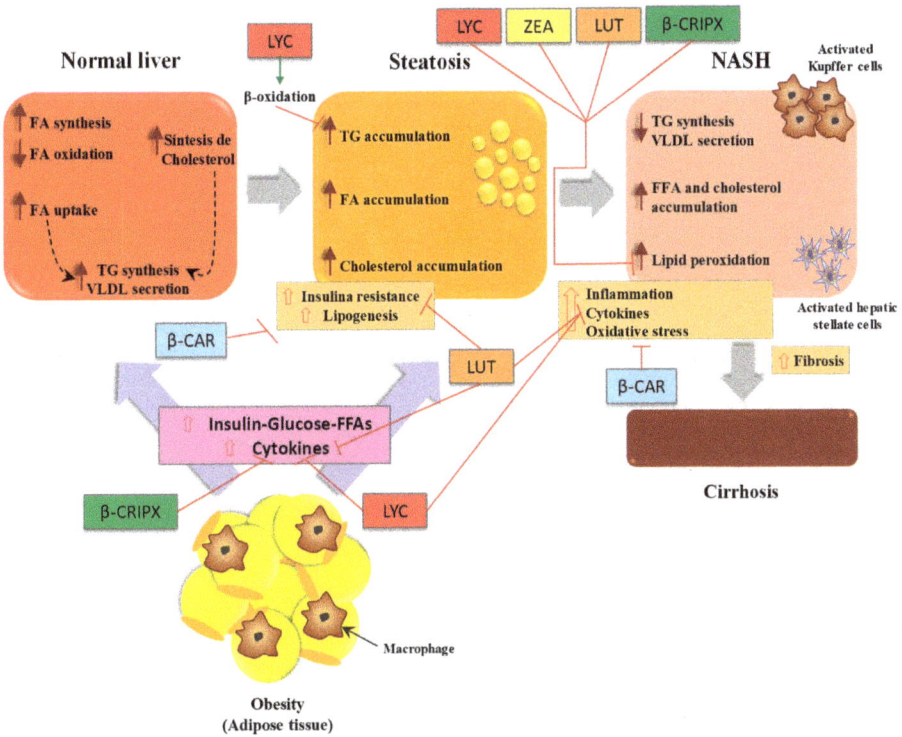

Figure 1. Diagram of the pathogenesis of non-alcoholic fatty liver disease (NAFLD) and the protective effect of carotenoids affecting different pathways. The red arrows denote blocked or decreased pathways, whereas the green arrows represent increased or promoted pathways. FA: fatty acids, TG: triglycerides, FFA: free fatty acids β-CAR: β-carotene, LYC: lycopene, LUT: lutein, β-CRIPX: β-cryptoxanthin, ZEA: zeaxanthin [19,110].

In addition, carotenoids in the diet, apart from being an important part of the antioxidant defense system, are also precursors of vitamin A, which can help rejuvenate the shape of hepatic stellate cells, preventing the progression of fibrosis to hepatocellular carcinoma [114]. A retrospective longitudinal study with 3336 middle-aged Chinese adults observed that the highest levels of serum carotenoids are associated with the improvement of the indicators of NAFLD, mediated by a reduction of the retinol binding protein 4 (*RBP4*), triglycerides, HOMA-IR, and body mass index [115]. The positive effects of carotenoids in the diet of preventing or treating NAFLD are their possible effects on hepatic health, individually, are taken into consideration and described below.

3.2. β-Carotene

This carotenoid has an important role as a precursor of vitamin A and has a direct impact on fighting against ROS, and hence protecting the body against oxidative stress [116,117]. Recent research has shown the possible preventive and protective effects of β-carotene on hepatic steatosis, fibrosis, oxidative stress, inflammation, and apoptosis [14]. In addition, this powerful antioxidant serves as a pre-hormone, since through metabolism, it is converted into retinoic acid, which functions as a ligand, regulating the expression of genes involved in metabolic processes [118].

Experimental studies have shown the potent hepatoprotective effect of β-carotene carried out in animal models, cell lines, and humans. Baybutt and Molteni [119] found that dietary supplementation of β-carotene had a protective effect on liver damage, demonstrating that rats with monocrotaline-induced steatosis decreased fat accumulation and liver hemorrhages. Patel and Sail [120] indicated that β-carotene protects physiological antioxidants against carcinogenesis induced by aflatoxin-B1 in albino rats. Another study with rats showed that supplementation with β-carotene increases the levels of vitamin C, glutathione, and enzymes related to glutathione, acting as scavengers of free radicals and consequently reducing the toxicity of aflatoxin-B1 [121]. In another animal study based on supplementation with (9Z)-β-carotene (isomer of β-carotene), a decrease of plasma cholesterol and atherogenesis index and a reduction of fat accumulation and inflammation was reported in the liver of mice fed a diet high in fat. This could be due to the transcriptional regulation of inflammatory cytokines, such as the vascular cell adhesion molecule 1 (*VCAM-1*), interleukin 1α (*IL-1α*), monocyte chemoattractant protein-1 (*MCP-1*), and interferon-γ (*INF- γ*) [122].

A study conducted by Ozturk et al. [123] found that the dietary intake of apricot reduced the risks of hepatic steatosis and the damage induced by carbon tetrachloride (CCl_4) in Wistar rats. Markers of oxidative stress, such as malondialdehyde (MDA), total levels of glutathione (GSH), catalase, superoxide dismutase (SOD), and GSH peroxidase activities (GSH-Px), were significantly altered by CCl_4. However, the liver damage and steatosis imposed by the high concentration of ROS were improved with the intake of apricots rich in β-carotene. Liu et al. [124] found that in a cell culture system, β-carotene could decrease the hepatosteatosis induced by the hepatitis C virus (HCV) by inhibiting RNA replication. Through its activity of provitamin A and its role in the inhibition of reactive oxygen species, β-carotene has been confirmed to have a positive effect on the progression of the hepatitis virus (HBV and HCV), preventing the development of carcinoma hepatocellular [125]. Another investigation showed that the Campari tomatoes, which contain more β-carotene and lycopene than normal tomatoes, improved diet-induced obesity, dyslipidemia, and hepatosteatosis through gene regulation related to lipogenesis in the model of zebrafish, transcriptionally lowering the expression of sterol regulatory element-binding transcription factor 1 (*SREBF1*) and increasing the expression of the forkhead box O1 gene (*FOXO1*) [126]. Other foods rich in β-carotene, such as goji berries (*Lycium barbarum*), have also improved liver fibrosis, oxidative stress, and inflammatory response in a rat model with NASH and cellular steatosis induced by a high-fat diet. These improvements were partially due to the modulation of the transcription factor NF-κB, the MAPK pathway, and the autophagic process [127].

A human investigation found that NAFLD has an inverse relationship with the nutritional status of vitamin A in individuals with class III obesity, observing low levels of retinol and serum β-carotene in patients with NAFLD, which entails a significant association between insulin resistance and retinol and β-carotene levels [128]. Moreover, a case-control study explored associations between dietary intake of vitamin A and carotenes (β-carotene), and the risk of primary liver cancer. They used a food frequency questionnaire to assess the usual dietary intake and through a logistic regression analysis, the researchers suggested that a higher dietary intake of retinol, carotene, and vitamin A (1000 µg RE/day) obtained from dietary sources was associated with a lower risk of primary liver cancer. The researchers also found that an intake of 2300 μg of RE/day of total vitamin A in the diet was the one with the lowest risk of primary liver cancer [129]. In addition, a recent study involving 62 patients with NAFLD and 24 control subjects showed that the serum levels of β-carotene and the ratio of β-carotene to retinol (SC/SR) in patients with NAFLD (hepatic steatosis, inflammation, and fibrosis) were significantly lower than in the controls. According to this, the researchers indicated that both β-carotene and SC/SR decreased gradually with the progression of the disease—from the normal liver, hepatic steatosis, to the limit of steatohepatitis. These results showed that a lower concentration of circulating β-carotene and an SC/SR ratio are associated with the histological severity of NAFLD [130].

3.3. Lycopene

The main protective effect of lycopene is due to its antioxidant effect through the inactivation of ROS and the extinction of free radicals [131]. Beyond its antioxidant capacity, there are many other potential non-antioxidant mechanisms, of which lycopene can protect against chronic diseases, including the regulation of gene expression, gap junctions, antiproliferative capacity, immune and hormonal modulation, among others [23,132,133]. This is why lycopene is one of the most studied carotenoids in the prevention and treatment of NAFLD [14].

It has been confirmed that this antioxidant has a potential hepatoprotective effect in hepatitis induced by D-galactosamine/lipopolysaccharide (D-GaIN/LPS) in rats, affecting the metabolism of lipoproteins, restoring the altered levels of lipid metabolizing enzymes, and stabilizing the disposition of lipoprotein levels [134,135]. A study conducted by Wang et al. [136] investigated the protective effect of the intake of lycopene and tomato extract in the hepatocarcinogenesis promoted by NASH in an in vivo study. In this study, Sprague-Dawley rats were used, and a single dose administration of diethylnitrosamine (DEN) was applied because of its origin in a hepatocellular carcinoma. Note that lycopene and tomato extract can inhibit hepatocarcinogenesis in relationships through the reduction of oxidative stress. In addition, a significant decrease in cytochrome P450 2E1, inflammatory foci, and mRNA expression of proinflammatory cytokines (*TNF-α*, *IL-1β*, and *IL-12*) were also found. A study conducted by Ahn et al. [137] indicated that lycopene altered the down regulation of the expression of miRNA-21 (*miR-21*) in mice, induced by a high-fat diet. As a regulator of gene expression at the posttranscriptional level, miR-21 was upregulated by the ingestion of lycopene, inhibiting the expression of the fatty acid-binding protein 7 (*FABP7*) and blocking the accumulation of intracellular lipids induced by stearic acid in Hepa 1–6 cells. It was indicated that lycopene prevented non-alcoholic steatohepatitis in rats and mice, which was induced by a high-fat diet, and a reduction in oxidative stress in cells was observed [138–140]. This demonstrates that the incorporation of this carotenoid in a balanced diet prevents NAFLD [16,141]. Kujawska et al. [142] noted that tomato paste intake in rats before administration of N-nitrosodiethylamine (NDEA) was effective in recovering the enzymes SOD, catalase, and glutathione reductase by 32%–97%, indicating the protective role against oxidative stress induced by NDEA. In addition, the study showed that DNA damage induced by NDEA in leukocytes decreased by 10% in rats treated with tomato paste. It has been suggested that lycopene supplementation in the diet prevents the incidence of hepatocellular carcinoma (HCC) induced by high-fat diets in mice, suppressing oncogenic signals, including methionine mRNA, β-catenin protein, and the activation of complex 1 of the target of rapamycin in mammalian cells (mTOR). This suggests that lycopene in the diet and its metabolites can be used in the prevention of liver cancer in patients with NAFLD [143,144]. Martín-Pozuelo et al. [145] studied the effect of tomato juice intake on gene expression in rats with induced hepatic steatosis, noting that supplementation with tomato juice led to an accumulation of all-*E* and *Z*-lycopene, as well as their metabolites in the liver of animals fed a normal diet + lycopene (NL) and a high-fat diet + lycopene (HL), with higher levels in the treatment of HL than in the NL group (63.07% vs. 44.45%) due to a higher absorption. In addition, it was shown that rats fed high-fat diets and tomato juice compared to rats that ingested only high-fat diets and water (NA) had significantly increased levels of high-density lipoproteins (HDL), and this also decreased oxidative stress through the reduction of isoprostanes in the urine. Regarding the analysis of gene expression of biomarkers associated with lipid metabolism, the overexpression of several genes related to the transport of fatty acids, lipid hydrolysis, and β-oxidation of mitochondrial and peroxisomal fatty acids was observed. In vitro and in vivo studies demonstrated that lycopene reduces ROS production in SK-Hep-1 cells by inhibiting dicotinamide adenine dinucleotide phosphate oxidase (NADPH) through protein kinase C (PKC) signaling. Furthermore, it was indicated that lycopene improved hepatotoxicity by acting as an antioxidant, mainly by reducing protein carbonylation and areas of necrosis, ameliorating the general appearance of the lesion in C57BL/6 mice [146]. A recent study showed that lycopene exerted anti-inflammatory activities against paracetamol liver injury (APAP) in C57BL/6 mice by improving the redox state [147]. Yefsah-Idres et al. [148] showed that rats with a diet high in methionine content had abnormal histological features accompanied by an increase in the levels of serum homocysteine, alanine aminotransferase (ALT), and aspartate aminotransferase (AST), as well as MDA hepatic and a decrease in the activities of cystathionine-β-synthase (CBS) and S-adenosyl-homocysteine hydrolase, indicating that lycopene supplementation reversed hyperhomocysteinemia (related to oxidative stress), providing additional evidence of the hepatoprotective effects of lycopene. Lycopene also showed beneficial effects against HCC by modulating cell proliferation, glycolysis, and ultrastructure of liver cells [149]. Xu et al. [150] confirmed that lycopene alleviates liver injury induced by aflatoxin B1 (AFB1) by improving hepatic oxidation and detoxification potential with Nrf2 activation. In another model of NAFLD and hypercholesterolemia induced by a high-fat diet, researchers showed that consumption of tomato juice had different effects depending on the diet. In the group of rats that were taking tomato juice (with and without steatosis), the genes involved in β-oxidation as well as the thrombospondin receptor (*CD36*) were positively regulated, and apolipoprotein B (*APOB*) and lipoprotein lipase (*LPL*) were negatively regulated. The accumulation of lycopene in rats with steatosis positively regulated the farnesoid X-activated receptor (*FXR*) and the hepatocyte nuclear factor 4 alpha (*HNF4A*), which have been suggested as preventive factors in relation to steatosis [151]. Regarding the metabolomic study, the intake of tomato juice in rats with fat-induced steatosis stimulated the biosynthesis of glutathione and the amino acids of the transulfurization pathway, increasing the levels of metabolites related to the antioxidant response [139,151].

3.4. Lutein

The property of this antioxidant is also based on the uptake of free radicals, especially singlet oxygen, protecting against oxidative damage [152], although an antiviral activity against hepatitis B has also been described, since it inhibits the transcription of the virus [153]. Kim et al. [154] observed that lutein (0.1 g/100 g for 12 weeks) decreased inflammation and oxidative stress in the liver and in the eyes of guinea pigs fed a hypercholesterolemic diet. This carotenoid could prevent the degenerative conditions of the liver by decreasing the accumulation of free cholesterol, attenuate lipid peroxidation (decreased MDA), and the production of proinflammatory cytokines (TNF-α). Furthermore, in this study, it was also observed that guinea pigs fed lutein also had a lower DNA-binding activity of NF-κB. These antioxidant effects suggest protective effects against NAFLD. A protective anticarcinogenic effect after NDEA induction in rats with HCC was described, noting that the administration of lutein inhibited carcinogenesis, probably due to the combination of its antioxidant activity and the activation of cytochrome P450 enzymes, as well as other detoxifying enzymes such as glutathione S-transferase (GST) and UDP-glucuronyl transferase [155]. Another study suggested that lutein supplementation may protect against hepatic lipid accumulation and insulin resistance induced by a high-fat diet. In addition, the study also investigated the effects of lutein on the expression of the peroxisome proliferator activated receptor (PPAR) because it plays an important role in lipid metabolism, finding that the high-fat diet significantly inhibited the expression of the PPAR, which was restored with lutein supplementation [156]. Murillo et al. [157] used a nanoemulsion of lutein (3.5 mg/day) in a hypercholesterolemic diet administered to guinea pigs during six weeks, and observed an increase in the concentrations of this carotenoid in plasma and liver, in addition to a point decrease in hepatic steatosis (24% lower as assessed histologically), total liver cholesterol, and plasma ALT activity. In addition, in the study, a 55% decrease in LDL was also found in the groups supplemented with lutein compared to the control groups. These results suggest the protective effects of this nanoemulsion on hepatic steatosis.

3.5. β-Cryptoxanthin

According to its antioxidant activity, plasma concentrations of β-cryptoxanthin are inversely related to oxidative DNA damage rates and lipid peroxidation [158]. In addition, in in vivo and in vitro studies, it was observed that β-cryptoxanthin has anti-inflammatory effects, modulating the immune response of macrophages [159]. Takayanagi et al. [160] demonstrated that oral administration of this carotenoid repressed the secretion of proinflammatory cytokines (TNF-α, IL-1, and IL-6) and improved lipid metabolism and energy consumption. Kobori et al. [161] demonstrated that β-cryptoxanthin improved dietary-induced NASH by suppressing the expression of inflammatory genes in mice. They observed that this carotenoid suppressed the expression of genes inducible by LPS and by TNF-α in NASH. The elevated levels of the thiobarbituric acid reactive substances (TBARS) of the oxidative stress marker were also reduced. Therefore, β-cryptoxanthin represses inflammation and the resulting fibrosis, probably by suppressing the increase and activation of macrophages/Kupffer cells, leukocytes, and T cells. In a model with mice (lipotoxic model), it was observed that β-cryptoxanthin reversed steatosis, inflammation, and progression of fibrosis in NASH, reversing insulin resistance and preventing steatohepatitis by decreasing the activation of macrophages or Kupffer cells [162]. Another study showed that supplementation with β-cryptoxanthin in patients with NAFLD inhibited the progression of this disease, suggesting that the intake of β-cryptoxanthin is very effective in elevating antioxidant and anti-inflammatory activities in patients with NAFLD (17).

3.6. Other Carotenoids

Other carotenoids, such as α-carotene and zeaxanthin, also showed beneficial effects against chronic liver injury. An investigation carried out by Murakoshi et al. [163] found that α-carotene had an inhibitory effect on spontaneous hepatic carcinogenesis in male mice, significantly decreasing the mean number of hepatomas. Zeaxanthin showed protective effects against NAFLD, decreasing oxidative stress and liver fibrosis, suggesting that the mechanism of action of zeaxanthin is related to its antioxidant capacity [164]. Epidemiological studies showed that zeaxanthin is inversely associated with the prevalence of NAFLD in Chinese populations of medium and advanced ages [165]. A summary of the studies examining the role of these carotenoids in chronic liver diseases is described in Table 3.

Table 3. Summary of studies in which carotenoids had a beneficial effect on chronic liver diseases in cell lines, and human and animal models.

Agent	Model	Main Results	Reference
β-carotene	Rat: carcinogenesis induced by AFB1	↑ Antioxidantes enzymes (GSH-Px, catalase, GST) and vitamin C ↓ Risk of toxicity due to AFB1	[120]
Alga *Dunaliella bardawil* (rich in (9Z)-β-carotene)	Mouse: fed high-fat diet, LDL receptor knockout mouse	↓ Plasma cholesterol and atherogenesis (VLDL y LDL) ↓ Accumulation of fat and liver inflammation ↓ Levels of hepatic inflammatory genes (VCAM-1, IL-1α, MCP-1, INF-γ)	[122]
Apricot (rich in β-carotene)	Rat: Hepatic steatosis and damage induced by CCL4	↓ Liver MDA ↑ Levels of total GSH, catalase, SOD and GSH-Px ↓ Oxidative stress ↓ Hepatic steatosis and liver damage	[124]
Tomato "Campari" (rich in β-carotene and lycopene)	Zebrafish: Obesity induced by diet	↓ *SREBF1* in the Marn ↑ *FOXO1* in the expression of genes ↓ Diet-induced obesity, dyslipidemia and hepatic steatosis	[127]
Lycium barbarum polysaccharides (rich in β-carotene)	Rat: NASH induced by a high-fat diet	↑ Modulation of NF-κB and the MAPK pathway ↓ Accumulation of liver fat, inflammatory liver response, fibrosis and oxidative stress ↑ Hepatoprotective properties	[128]
Dietary carotenes and vitamin A	Human: patients with primary liver cancer	↓ Risk of primary liver cancer	[129]
Lycopene	Rat: NASH induced by high-fat diet	↓ Levels of CYP2E1 protein, MDA (plasma and liver) and TNF-α ↑ Hepatic GSH level ↓ Steatosis and inflammation	[138]
Tomato juice	Rat: hypercholesterolemic and NAFLD induced by the diet	↓ Levels of TG in plasma and isoprostanes in urine ↑ Accumulation of lycopene in the liver ↑ Relief of amino acid depletion ↑ Recovery of the redox balance in the liver ↑ Levels of L-carnitine ↑ Protective effect of NAFLD	[139]
Tomato juice	Rat: NAFLD induced by a high-fat diet	↓ Isoprostanes in urine, plasma TG and LDL ↑ Activity of mitochondrial β-oxidation and peroxisomal ↓ Steatosis	[145]

Table 3. *Cont.*

Agent	Model	Main Results	Reference
Lycopene	SK-Hep-1 cells: PKC pathway mediated by ROS production. Mouse: Hepatotoxicity induced by APAP overdose	↓ Production of ROS, NADPH oxidase and *MMP-2*, GSSG ↑ GSH and CAT	[146]
Lycopene	Rat: NAFLD induced by a high-fat diet	↓ ALT, AST, triglyceride, total cholesterol, MDA, LDL and FFA ↓ *CYP2E1* and *TNF-α* ↑ GSH, SOD y HDL ↑ Protective effect on NAFLD	[16]
Lycopene	Rat: NAFLD induced by a high-fat diet	↓ liver weight, LDL and liver total cholesterol ↑ GSH-Px, SOD and CAT en the liver	[141]
Lycopene	Mouse: liver injury induced by AFB1	↓ Acummulatio of AFB1-ADN adducts in the liver ↑ Activation of Nrf2 signaling ↑ Antioxidant potential and liver detoxification	[150]
Tomato juice	Rat: hypercholesterolemic and NAFLD induced by the diet	↑ Regulation of *CD36*, *FXR* and *HNF4A* ↓ Regulation of *APOB* and *LPL* ↓ Synthesis of fatty acids, triglycerides and cholesterol ↑ Levels of metabolites related to the antioxidant response	[151]
Lutein	Guinea pig: Hepatic steatosis induced by a hypercholesterolemic diet	↓ Hepatic free cholesterol ↓ Malondialdehyde and hepatic TNF-α ↓ Binding to the hepatic DNA of NF-κB	[154]
Lutein	Rat: Hepatocellular carcinoma induced by *N*-nitrosodiethylamine (NDEA)	↓ ALT, AST, alkaline phosphatase in plasma and liver tissue ↑ GSH ↓ GGT ↑ UDP-glucoronyl transferase and glutathione-S-transferase	[155]
Lutein	Rat: NAFLD induced by a high-fat diet	↓ Liver total cholesterol and triglycerides ↑ HDL in serum ↓ ALT in serum ↑ Hepatic insulin sensitivity ↑ Hepatic fatty acids catabolism	[156]
Lutein	Guinea pig: Hepatic steatosis induced by a hypercholesterolemic diet	↓ Hepatic steatosis (evaluated histologically) ↓ Total hepatic cholesterol ↓ Plasma ALT and LDL activity	[157]

Table 3. *Cont.*

Agent	Model	Main Results	Reference
β-cryptoxanthin	Mouse: Obese model	↓ Body weight and abdominal adipose tissue ↓ Triglycerides and serum total cholesterol ↓ Inflammatory citokines ↑ Lipid metabolism and energy consumption	[160]
β-cryptoxanthin	Mouse: NASH induced by a diet high in cholesterol and high in fat	↓ Liver TBARS ↑ Suppresses the expression of the inducible *LPS* and *TNF-α* genes ↓ Inflammatory response (suppresses the activation of macrophages, T helper and citotoxic cells)	[161]
β-cryptoxanthin	Mouse: Hepatic steatosis and NASH induced by the diet high in fat and cholesterol	↓ Total content of hepatic macrophages and T cells	[162]
β-cryptoxanthin	Human: Patients with NAFLD (NASH and NAFL)	↓ GGT, LDL and serum IL-6 ↑ SOD and serum IL-10 ↑ Antioxidant and anti-inflammtory activities	[17]
α-carotene	Mouse: spontaneous hepatic carcinogenesis	↓ Hepatomas	[163]
Zeaxanthin	Gerbil from Mongolia: NASH induced by a diet deficient in methionine and choline	↓ Liver fibrosis ↓ Hepatic lipid hydroperoxides	[164]

4. Conclusions

The available evidence regarding the potential use of dietary carotenoids in liver health suggests that these compounds are effective in reducing lipid accumulation, insulin resistance, oxidative stress, and inflammation of hepatocytes, which is why they could be used as a dietary alternative for the prevention and treatment of NAFLD. The effects of the specific mechanisms by which carotenoids protect against NAFLD are depicted in Figure 1. The antioxidant and anti-inflammatory properties are the main mechanisms of action of carotenoids, modulating intracellular signaling pathways that influence gene expression and protein translation. During the last decade, several investigations were carried out, attempting to elucidate the protective function of carotenoids against the lesions induced by oxidative stress, especially the anticancer effects, by preventing tumors and modulating the proliferation of liver cells. In addition, several carotenoids have provitamin A activity, which helps to rejuvenate the shape of hepatic stellate cells and prevent the progression of fibrosis to HCC. More preclinical and clinical studies are needed to evaluate if there could be an effective dose, considering the bioavailability of dietary carotenoids, for the prevention of NAFLD.

Author Contributions: Writing—original draft preparation, L.I.E.-T. and M.J.P.-C.; writing—review and editing L.I.E.-T., M.J.P.-C. and J.G.-A. All authors read and approved the final manuscript.

Funding: This research was funding by the project Ref. No. 20904/PI/18 from "Fundación Séneca", Regional Agency of Research of "Comunidad Autónoma de la Región de Murcia".

Acknowledgments: Laura Inés Elvira-Torales thanks the Mexican Public Education Secretary for a Doctoral Scholarship (ITESTB-003 PRODEP Program).

Conflicts of Interest: The authors declare that there was no conflict of interest. All authors have accepted their publication in *Antioxidants*.

Abbreviations

MAPK	Mitogen-activated protein kinase
CYP2E1	Cytochrome P450 family 2 subfamily E member 1
LDL	Low density lipoproteins
HDL	High density lipoprotein
Nrf2	Nuclear factor erythroid 2-related factor 2
GGT	Gamma-glutamyltransferase
UDP	Uridine diphosphate
IL-6	Interleukin-6
IL-10	Interleukin-10
NF-κB	Nuclear factor kappa B

References

1. Maiani, G.; Periago Castón, M.J.; Catasta, G.; Toti, E.; Cambrodón, I.G.; Bysted, A.; Böhm, V. Carotenoids: Actual knowledge on food sources, intakes, stability and bioavailability and their protective role in humans. *Mol. Nutr. Food Res.* **2009**, *53*, S194–S218. [CrossRef] [PubMed]
2. Sugiura, M. Carotenoids: Liver diseases prevention. In *Bioactive Foods as Dietary Interventions for Liver and Gastrointestinal Disease*, 1st ed.; Watson, R.R., Preedy, V.R., Eds.; Academic Press: San Diego, CA, USA, 2013; pp. 421–436, ISBN 9780123971548.
3. Coyne, T.; Ibiebele, T.I.; Baade, P.D.; McClintock, C.S.; Shaw, J.E. Metabolic syndrome and serum carotenoids: Findings of a cross-sectional study in Queensland, Australia. *Br. J. Nutr.* **2009**, *102*, 1668–1677. [CrossRef] [PubMed]
4. Sharoni, Y.; Linnewiel-Hermoni, K.; Khanin, M.; Salman, H.; Veprik, A.; Danilenko, M.; Levy, J. Carotenoids and apocarotenoids in cellular signaling related to cancer: A review. *Mol. Nutr. Food Res.* **2012**, *56*, 259–269. [CrossRef] [PubMed]

5. Periago, M.J.; García-Alonso, J. Biodisponibilidad de antioxidantes en la dieta. In *Antioxidantes en Alimentos y Salud*, 1st ed.; Álvarez-Parilla, E., González-Aguilar, A., De la Rosa, L.A., Ayala-Zavala, J.F., Eds.; AM-Editores: Ciudad de Mexico, Mexico, 2012; pp. 257–291, ISBN 9786074372076.
6. Bonet, M.L.; Canas, J.A.; Ribot, J.; Palou, A. Carotenoids in adipose tissue biology and obesity. *Subcell. Biochem.* **2016**, *79*, 377–414. [PubMed]
7. Darvin, M.E.; Sterry, W.; Lademann, J.; Vergou, T. The role of carotenoids in human skin. *Molecules* **2011**, *16*, 10491–10506. [CrossRef]
8. Meinke, M.C.; Darvin, M.E.; Vollert, H.; Lademann, J. Bioavailability of natural carotenoids in human skin compared to blood. *Eur. J. Pharm. Biopharm.* **2010**, *76*, 269–274. [CrossRef] [PubMed]
9. Evans, J.A.; Johnson, E.J. The role of phytonutrients in skin health. *Nutrients* **2010**, *2*, 903–928. [CrossRef] [PubMed]
10. Mikolasevic, I.; Milic, S.; Turk Wensveen, T.; Grgic, I.; Jakopcic, I.; Stimac, D.; Orlic, L. Nonalcoholic fatty liver disease—A multisystem disease? *World J. Gastroenterol.* **2016**, *22*, 9488–9505. [CrossRef]
11. Azzam, H.; Malnick, S. Non-alcoholic fatty liver disease—The heart of the matter. *World J. Hepatol.* **2015**, *7*, 1369–1376. [CrossRef]
12. Ferramosca, A.; Di Giacomo, M.; Zara, V. Antioxidant dietary approach in treatment of fatty liver: New insights and updates. *World J. Gastroenterol.* **2017**, *23*, 4146–4157. [CrossRef]
13. Murillo, A.G.; DiMarco, D.M.; Fernandez, M.L. The potential of non-provitamin A carotenoids for the prevention and treatment of non-alcoholic fatty liver disease. *Biology* **2016**, *5*, 42. [CrossRef] [PubMed]
14. Yilmaz, B.; Sahin, K.; Bilen, H.; Bahcecioglu, I.H.; Bilir, B.; Ashraf, S.; Kucuk, O. Carotenoids and non-alcoholic fatty liver disease. *Hepatobiliary Surg. Nutr.* **2015**, *4*, 161–171. [PubMed]
15. Christensen, K.; Lawler, T.; Mares, J. Dietary carotenoids and non-alcoholic fatty liver disease among US adults, NHANES 2003–2014. *Nutrients* **2019**, *11*, 1101. [CrossRef] [PubMed]
16. Jiang, W.; Guo, M.H.; Hai, X. Hepatoprotective and antioxidant effects of lycopene on non-alcoholic fatty liver disease in rat. *World J. Gastroenterol.* **2016**, *22*, 10180–10188. [CrossRef] [PubMed]
17. Matsuura, B.; Miyake, T.; Yamamoto, S.; Furukawa, S.; Hiasa, Y. Usefulness of Beta-cryptoxanthin for nonalcoholic fatty liver diseases. *J. Food Nutr. Disord.* **2016**, *5*, 3.
18. Elvira-Torales, L.I.; Martín-Pozuelo, G.; González-Barrio, R.; Navarro-González, I.; Pallarés, F.J.; Santaella, M.; Periago-Castón, M.J. Ameliorative effect of spinach on non-alcoholic fatty liver disease induced in rats by a high-fat diet. *Int. J. Mol. Sci.* **2019**, *20*, 1662. [CrossRef] [PubMed]
19. Kitade, H.; Chen, G.; Ni, Y.; Ota, T. Nonalcoholic fatty liver disease and insulin resistance: New insights and potential new treatments. *Nutrients* **2017**, *9*, 387. [CrossRef]
20. Milani, A.; Basirnejad, M.; Shahbazi, S.; Bolhassani, A. Carotenoids: Biochemistry, pharmacology and treatment. *Br. J. Pharmacol.* **2017**, *174*, 1290–1324. [CrossRef]
21. Johnson, E.J. The role of carotenoids in human health. *Nutr. Clin. Care* **2002**, *5*, 56–65. [CrossRef]
22. Ruiz-Sola, M.Á.; Rodríguez-Concepción, M. Carotenoid biosynthesis in Arabidopsis: A colorful pathway. *Arabidopsis Book* **2012**, *10*, e0158. [CrossRef]
23. Rao, A.V.; Rao, L.G. Carotenoids and human health. *Pharmacol. Res.* **2007**, *55*, 207–216. [CrossRef]
24. Stahl, W.; Sies, H. Antioxidant activity of carotenoids. *Mol. Aspects Med.* **2003**, *24*, 345–351. [CrossRef]
25. Stahl, W.; Sies, H. Bioactivity and protective effects of natural carotenoids. *Biochim. Biophys. Acta* **2005**, *1740*, 101–107. [CrossRef]
26. Saini, R.K.; Nile, S.H.; Park, S.W. Carotenoids from fruits and vegetables: Chemistry, analysis, occurrence, bioavailability and biological activities. *Food Res. Int.* **2015**, *76*, 735–750. [CrossRef]
27. Eggersdorfer, M.; Wyss, A. Carotenoids in human nutrition and health. *Arch. Biochem. Biophys.* **2018**, *652*, 18–26. [CrossRef]
28. Bohn, T. Bioavailability of non-provitamin A carotenoids. *Curr. Nutr. Food Sci.* **2008**, *4*, 240–258. [CrossRef]
29. Ornelas-Paz, J.J.; Yahia, E.M.; Gadea-Béjar, A.A.; Pérez-Martínez, J.D.; Ochoa-Reyes, E. Biodisponibilidad y actividad biológica de carotenoides y vitamina A. In *Antioxidantes en Alimentos y Salud*, 1st ed.; Álvarez-Parilla, E., González-Aguilar, A., De la Rosa, L.A., Ayala-Zavala, J.F., Eds.; AM-Editores: Ciudad de Mexico, Mexico, 2012; pp. 293–327, ISBN 9786074372076.
30. Von Elbe, J.H.; Schwartz, S.J. Colorantes. In *Química de Los Alimentos*, 2nd ed.; Fennema, O.R., Ed.; ACRIBIA: Zaragoza, Spain, 2000; pp. 773–850, ISBN 8420009148.

31. Rühl, R. Non-pro-vitamin A and pro-vitamin A carotenoids in atopy development. *Int. Arch. Allergy Immunol.* **2013**, *161*, 99–115. [CrossRef]

32. Woodside, J.V.; McGrath, A.J.; Lyner, N.; McKinley, M.C. Carotenoids and health in older people. *Maturitas* **2015**, *80*, 63–68. [CrossRef]

33. Khachik, F.; Sprangler, C.J.; Smith, J.C.; Canfield, L.M.; Steck, A.; Pfander, H. Identification, quantification, and relative concentrations of carotenoids and their metabolites in human milk and serum. *Anal. Chem.* **1997**, *69*, 1873–1881.

34. Carlsen, M.H.; Karlsen, A.; Lillegaard, I.T.; Gran, J.M.; Drevon, C.A.; Blomhoff, R.; Andersen, L.F. Relative validity of fruit and vegetable intake estimated from an FFQ, using carotenoid and flavonoid biomarkers and the method of triads. *Br. J. Nutr.* **2011**, *105*, 1530–1538. [CrossRef]

35. Baldrick, F.R.; Woodside, J.V.; Elborn, J.S.; Young, I.S.; McKinley, M.C. Biomarkers of fruit and vegetable intake in human intervention studies: A systematic review. *Crit. Rev. Food Sci. Nutr.* **2011**, *51*, 795–815. [CrossRef]

36. Britton, G.; Khachik, F. Carotenoids in Food. In *Carotenoids: Nutrition and Health*, 4th ed.; Britton, G., Pfander, H., Liaaen-Jensen, S., Eds.; Birkhäuser Verlag: Basel, Switzerland, 2009; pp. 45–66.

37. Shete, V.; Quadro, L. Mammalian metabolism of β-carotene: Gaps in knowledge. *Nutrients* **2013**, *5*, 4849–4868. [CrossRef]

38. Rodriguez-Amaya, D.B.; Kimura, M.; Godoy, H.T.; Amaya-Farfan, J. Updated brazilian database on food carotenoids: Factors affecting carotenoids composition. *J. Food Compos. Anal.* **2008**, *21*, 445–463. [CrossRef]

39. Latief, U.; Ahmad, R. Role of dietary carotenoids in different etiologies of chronic liver diseases. In *Descriptive Food Science*, 1st ed.; Valero Díaz, A., García-Gimeno, R.M., Eds.; IntechOpen: London, UK, 2018; pp. 93–112, ISBN 9781789845952.

40. O'Neill, M.E.; Carroll, Y.; Corridan, B.; Olmedilla, B.; Granado, F.; Blanco, I.; Southon, S. A European carotenoid database to assess carotenoid intakes and its use in a five-country comparative study. *Br. J. Nutr.* **2001**, *85*, 499–507. [CrossRef]

41. Perry, A.; Rasmussen, H.; Johnson, E.J. Xanthophyll (lutein, zeaxanthin) content in fruits, vegetables and corn and egg products. *J. Food Compos. Anal.* **2009**, *22*, 9–15. [CrossRef]

42. Šivel, M.; Kráčmar, S.; Fišera, M.; Klejdus, B.; Kubáň, V. Lutein content in marigold flower (*Tagetes erecta* L.) concentrates used for production of food supplements. *Czech J. Food Sci.* **2014**, *32*, 521–525. [CrossRef]

43. González-Barrio, R.; Periago, M.J.; Luna-Recio, C.; Garcia-Alonso, F.J.; Navarro-González, I. Chemical composition of the edible flowers, pansy (*Viola wittrockiana*) and snapdragon (*Antirrhinum majus*) as new sources of bioactive compounds. *Food Chem.* **2018**, *252*, 373–380. [CrossRef]

44. Gammone, M.A.; Riccioni, G.; D'Orazio, N. Carotenoids: Potential allies of cardiovascular health? *Food Nutr. Res.* **2015**, *59*, 26762.

45. Burri, B.J.; La Frano, M.R.; Zhu, C. Absorption, metabolism, and functions of β-cryptoxanthin. *Nutr. Rev.* **2016**, *74*, 69–82. [CrossRef]

46. Krinsky, N.I.; Johnson, E.J. Carotenoid actions and their relation to health and disease. *Mol. Aspects Med.* **2005**, *26*, 459–516. [CrossRef]

47. Dias, M.G.; Olmedilla-Alonso, B.; Hornero-Méndez, D.; Mercadante, A.Z.; Osorio, C.; Vargas-Murga, L.; Meléndez-Martínez, A.J. Comprehensive database of carotenoid contents in Ibero-American foods. A valuable tool in the context of functional foods and the establishment of recommended intakes of bioactives. *J. Agric. Food Chem.* **2018**, *66*, 5055–5107. [CrossRef]

48. Giuffrida, D.; Torre, G.; Dugo, P.; Dugo, G. Determination of the carotenoid profile in peach fruits, juice and jam. *Fruits* **2012**, *68*, 39–44. [CrossRef]

49. Mercadante, A.Z. Carotenoids in foods: Sources and stability during processing and storage. In *Food Colorants: Chemical and Functional Properties*, 1st ed.; Socaciu, C., Ed.; CRC Press: Boca Raton, FL, USA, 2008; p. 213, ISBN 9780849393570.

50. Yahia, E.M.; Ornelas-Paz, J.J. Chemistry, stability, and biological actions of carotenoids. In *Fruit and Vegetable Phytochemicals: Chemistry, Nutritional Value, and Stability*, 1st ed.; de la Rosa, L.A., Álvarez-Parrilla, E., González-Aguilar, G.A., Eds.; Wiley-Blackwell: Ames, IA, USA, 2010; pp. 177–222, ISBN 978-0-813-80320-3.

51. Boon, C.S.; McClements, D.J.; Weiss, J.; Decker, E.A. Factors influencing the chemical stability of carotenoids in foods. *Crit. Rev. Food Sci. Nutr.* **2010**, *50*, 515–532. [CrossRef]

52. Marze, S. Bioaccessibility of lipophilic micro-constituents from a lipid emulsion. *Food Funct.* **2015**, *6*, 3218–3227. [CrossRef]

53. Van Het Hof, K.H.; West, C.E.; Weststrate, J.A.; Hautvast, J.G. Dietary factors that affect the bioavailability of carotenoids. *J. Nutr.* **2000**, *130*, 503–506. [CrossRef]

54. Donhowe, E.G.; Kong, F. Beta-carotene: Digestion, microencapsulation, and in vitro bioavailability. *Food Bioprocess Technol.* **2014**, *7*, 338–354. [CrossRef]

55. Rein, M.J.; Renouf, M.; Cruz-Hernandez, C.; Actis-Goretta, L.; Thakkar, S.K.; Da Silva Pinto, M.; da Silva Pinto, M. Bioavailability of bioactive food compounds: A challenging journey to bioefficacy. *Br. J. Clin. Pharmacol.* **2013**, *75*, 588–602. [CrossRef]

56. Colle, I.J.; Lemmens, L.; Knockaert, G.; Van Loey, A.; Hendrickx, M. Carotene Degradation and Isomerization during Thermal Processing: A Review on the Kinetic Aspects. *Crit. Rev. Food Sci. Nutr.* **2016**, *56*, 1844–1855. [CrossRef]

57. Lemmens, L.; Colle, I.; Van Buggenhout, S.; Palmero, P.; Van Loey, A.; Hendrickx, M. Carotenoid bioaccessibility in fruit- and vegetable-based food products as affected by product (micro)structural characteristics and the presence of lipids: A review. *Trends Food Sci. Technol.* **2014**, *38*, 125–135. [CrossRef]

58. Reboul, E. Absorption of vitamin A and carotenoids by the enterocyte: Focus on transport proteins. *Nutrients* **2013**, *5*, 3563–3581. [CrossRef]

59. Desmarchelier, C.; Borel, P. Overview of carotenoid bioavailability determinants: From dietary factors to host genetic variations. *Trends Food Sci. Technol.* **2017**, *69*, 270–280. [CrossRef]

60. Priyadarshani, A.M.B. A review on factors influencing bioaccessibility and bioefficacy of carotenoids. *Crit. Rev. Food Sci. Nutr.* **2017**, *57*, 1710–1717. [CrossRef]

61. Bernhardt, S.; Schlich, E. Impact of different cooking methods on food quality: Retention of lipophilic vitamins in fresh and frozen vegetables. *J. Food Eng.* **2006**, *77*, 327–333. [CrossRef]

62. Fernandez-Garcia, E.; Carvajal-Lerida, I.; Jaren-Galan, M.; Garrido-Fernandez, J.; Perez-Galvez, A.; Hornero-Mendez, D. Carotenoids bioavailability from foods: From plant pigments to efficient biological activities. *Food Res. Int.* **2012**, *46*, 438–450. [CrossRef]

63. Honda, M.; Maeda, H.; Fukaya, T.; Goto, M. Effects of Z-isomerization on the bioavailability and functionality of carotenoids: A review. In *Descriptive Food Science*, 1st ed.; Valero Díaz, A., García-Gimeno, R.M., Eds.; IntechOpen: London, UK, 2018; pp. 141–159, ISBN 9781789845952.

64. Cooperstone, J.L.; Ralston, R.A.; Riedl, K.M.; Haufe, T.C.; Schweiggert, R.M.; King, S.A.; Schwartz, S.J. Enhanced bioavailability of lycopene when consumed as *cis*-isomers from *tangerine* compared to red tomato juice, a randomized, cross-over clinical trial. *Mol. Nutr. Food Res.* **2015**, *59*, 658–669. [CrossRef]

65. Coral-Hinostroza, G.N.; Ytrestøyl, T.; Ruyter, B.; Bjerkeng, B. Plasma appearance of unesterified astaxanthin geometrical *E/Z* and optical *R/S* isomers in men given single doses of a mixture of optical 3 and 3′R/S isomers of astaxanthin fatty acyl diesters. *Comp. Biochem. Physiol. Part C Toxicol. Pharmacol.* **2004**, *139*, 99–110. [CrossRef]

66. Honda, M.; Kageyama, H.; Hibino, T.; Zhang, Y.; Diono, W.; Kanda, H.; Goto, M. Improved carotenoid processing with sustainable solvents utilizing Z-isomerization-induced alteration in physicochemical properties: A review and future directions. *Molecules* **2019**, *7*, 2149. [CrossRef]

67. Palafox-Carlos, H.; Ayala-Zavala, J.F.; González-Aguilar, G.A. The role of dietary fiber in the bioaccessibility and bioavailability of fruit and vegetable antioxidants. *J. Food Sci.* **2011**, *76*, R6–R15. [CrossRef]

68. Yeum, K.J.; Russell, R.M. Carotenoid bioavailability and bioconversion. *Annu. Rev. Nutr.* **2002**, *22*, 483–504. [CrossRef]

69. Goltz, S.R.; Campbell, W.W.; Chitchumroonchokchai, C.; Failla, M.L.; Ferruzzi, M.G. Meal triacylglycerol profile modulates postprandial absorption of carotenoids in humans. *Mol. Nutr. Food Res.* **2012**, *56*, 866–877. [CrossRef]

70. Roodenburg, A.J.; Leenen, R.; Hof, K.H.; Weststrate, J.A.; Tijburg, L.B. Amount of fat in the diet affects bioavailability of lutein esters but not of alpha-carotene, beta-carotene, and vitamin E in humans. *Am. J. Clin. Nutr.* **2000**, *71*, 1187–1193. [CrossRef]

71. Periago, M.J.; Bravo, S.; García-Alonso, F.J.; Rincón, F. Detection of key factors affecting lycopene in vitro accessibility. *J. Agric. Food Chem.* **2013**, *61*, 3859–3867. [CrossRef]

72. Lakshminarayana, R.; Raju, M.; Prakash, M.K.; Baskaran, V. Phospholipid, oleic acid micelles and dietary olive oil influence the lutein absorption and activity of antioxidant enzymes in rats. *Lipids* **2009**, *44*, 799–806. [CrossRef]

73. Victoria-Campos, C.I.; Ornelas-Paz, J.; De, J.; Yahia, E.M.; Failla, M.L. Effect of the interaction of heat-processing style and fat type on the micellarization of lipid-soluble pigments from green and red pungent peppers (*Capsicum annuum*). *J. Agric. Food Chem.* **2013**, *61*, 3642–3653. [CrossRef]

74. Hoffmann, J.; Linseisen, J.; Riedl, J.; Wolfram, G. Dietary fiber reduces the antioxidative effect of a carotenoid and alpha-tocopherol mixture on LDL oxidation ex vivo in humans. *Eur. J. Nutr.* **1999**, *38*, 278–285. [CrossRef]

75. Soukoulis, C.; Bohn, T. A comprehensive overview on the micro- and nano-technological encapsulation advances for enhancing the chemical stability and bioavailability of carotenoids. *Crit. Rev. Food Sci. Nutr.* **2018**, *58*, 1–36. [CrossRef]

76. Fiedor, J.; Burda, K. Potential role of carotenoids as antioxidants in human health and disease. *Nutrients* **2014**, *6*, 466–488. [CrossRef]

77. Jansen, M.C.; Van Kappel, A.L.; Ocké, M.C.; Van 't Veer, P.; Boshuizen, H.C.; Riboli, E.; Bueno-de-Mesquita, H.B. Plasma carotenoid levels in Dutch men and women, and the relation with vegetable and fruit consumption. *Eur. J. Clin. Nutr.* **2004**, *58*, 1386–1395. [CrossRef]

78. Brevik, A.; Andersen, L.F.; Karlsen, A.; Trygg, K.U.; Blomhoff, R.; Drevon, C.A. Six carotenoids in plasma used to assess recommended intake of fruits and vegetables in a controlled feeding study. *Eur. J. Clin. Nutr.* **2004**, *58*, 1166–1173. [CrossRef]

79. Bohn, T.; McDougall, G.J.; Alegría, A.; Alminger, M.; Arrigoni, E.; Aura, A.M.; Martínez-Cuesta, M.C. Mind the gap-deficits in our knowledge of aspects impacting the bioavailability of phytochemicals and their metabolites–a position paper focusing on carotenoids and polyphenols. *Mol. Nutr. Food Res.* **2015**, *59*, 1307–1323. [CrossRef]

80. Bohn, T.; Desmarchelier, C.; Dragsted, L.O.; Nielsen, C.S.; Stahl, W.; Rühl, R.; Borel, P. Host-related factors explaining interindividual variability of carotenoid bioavailability and tissue concentrations in humans. *Mol. Nutr. Food Res.* **2017**, *61*, 1600685. [CrossRef]

81. Riboli, E.; Péquignot, G.; Repetto, F.; Axerio, M.; Raymond, L.; Boffetta, P.; Tuyns, A.J. A comparative study of smoking, drinking and dietary habits in population samples in France, Italy, Spain and Switzerland. I. Study design and dietary habits. *Rev. Epidemiol. Sante Publique* **1988**, *36*, 151–165.

82. Olmedilla, B.; Granado, F.; Southon, S.; Wright, A.J.; Blanco, I.; Gil-Martinez, E.; Thurnham, D.I. Serum concentrations of carotenoids and vitamins A, E, and C in control subjects from five European countries. *Br. J. Nutr.* **2001**, *85*, 227–238. [CrossRef]

83. Elia, M.; Stratton, R.J. Geographical inequalities in nutrient and risk of malnutrition among English people aged 65 y and older. *Nutrition* **2005**, *21*, 1100–1106. [CrossRef]

84. Rodriguez-Concepcion, M.; Avalos, J.; Bonet, M.L.; Boronat, A.; Gomez-Gomez, L.; Hornero-Mendez, D.; Ribot, J. A global perspective on carotenoids: Metabolism, biotechnology, and benefits for nutrition and health. *Prog. Lipid Res.* **2018**, *70*, 62–93. [CrossRef]

85. Stice, C.P.; Xia, H.; Wang, X.D. Tomato lycopene prevention of alcoholic fatty liver disease and hepatocellular carcinoma development. *Chronic Dis. Transl. Med.* **2018**, *4*, 211–224. [CrossRef]

86. Lucarini, M.; Lanzi, S.; D'Evoli, L.; Aguzzi, A.; Lombardi-Boccia, G. Intake of vitamin A and carotenoids from the Italian population–results of an Italian total diet study. *Int. J. Vitam. Nutr. Res.* **2006**, *76*, 103–109. [CrossRef]

87. Omenn, G.S.; Goodman, G.E.; Thornquist, M.D.; Balmes, J.; Cullen, M.R.; Glass, A.; Barnhart, S. Risk factors for lung cancer and for intervention effects in CARET, the Beta-Carotene and Retinol Efficacy Trial. *J. Natl. Cancer Inst.* **1996**, *88*, 1550–1559. [CrossRef]

88. Blumberg, J.; Block, G. The Alpha-Tocopherol, Beta-Carotene cancer prevention study in Finland. *Nutr. Rev.* **1994**, *52*, 242–245. [CrossRef]

89. Haider, C.; Ferk, F.; Bojaxhi, E.; Martano, G.; Stutz, H.; Bresgen, N.; Eckl, P. Effects of β-carotene and its cleavage products in primary pneumocyte type II cells. *Antioxidants* **2017**, *6*, 37. [CrossRef]

90. Ranard, K.M.; Jeon, S.; Mohn, E.S.; Griffiths, J.C.; Johnson, E.J.; Erdman, J.W., Jr. Dietary guidance for lutein: Consideration for intake recommendations is scientifically supported. *Eur. J. Nutr.* **2017**, *56*, 37–42. [CrossRef]

91. Grune, T.; Lietz, G.; Palou, A.; Ross, A.C.; Stahl, W.; Tang, G.; Biesalski, H.K. Beta-carotene is an important vitamin A source for humans. *J. Nutr.* **2010**, *140*, 2268S–2285S. [CrossRef]

92. Sánchez-Moreno, C.; Cano, M.P.; De Ancos, B.; Plaza, L.; Olmedilla, B.; Granado, F.; Martín, A. Mediterranean vegetable soup consumption increases plasma vitamin C and decreases F2-isoprostanes, prostaglandin E2 and monocyte chemotactic protein-1 in healthy humans. *J. Nutr. Biochem.* **2006**, *17*, 183–189. [CrossRef]

93. Heath, E.; Seren, S.; Sahin, K.; Kucuk, O. The role of tomato lycopene in the treatment of prostate cancer. In *Tomatoes, Lycopene and Human Health: Preventing Chronic Diseases*, 1st ed.; Rao, A.V., Ed.; Caledonian Science Press: Scotland, UK, 2007; pp. 127–140, ISBN 9780955356506.

94. Huang, Y.M.; Dou, H.L.; Huang, F.F.; Xu, X.R.; Zou, Z.Y.; Lin, X.M. Effect of supplemental lutein and zeaxanthin on serum, macular pigmentation, and visual performance in patients with early age-related macular degeneration. *BioMed Res. Int.* **2015**, *2015*, 564738. [CrossRef]

95. Gyamfi, D.; Patel, V. Liver metabolism: Biochemical and molecular regulations. In *Nutrition, Diet Therapy, and the Liver*, 1st ed.; Preedy, V.R., Lakshman, R., Srirajaskanthan, R., Watson, R.R., Eds.; CRC Press: Boca Raton, FL, USA, 2017; pp. 3–15, ISBN 9781138111790.

96. Cabré Gelada, E.; Peña Quintana, L.; Virgili Casas, N. Nutrición en las enfermedades hepatobiliares. In *Tratado de Nutrición: Nutrición y Enfermedad*, 3rd ed.; Gil, A., Ed.; Editorial Médica Panamericana: Madrid, Spain, 2017; pp. 865–906, ISBN 9788491101949.

97. Fabbrini, E.; Sullivan, S.; Klein, S. Obesity and nonalcoholic fatty liver disease: Biochemical, metabolic, and clinical implications. *Hepatology* **2010**, *51*, 679–689. [CrossRef]

98. Singhal, S.; Baker, S.S.; Baker, R.D.; Zhu, L. Role of paraoxonase 1 as an antioxidant in nonalcoholic steatohepatitis. In *The Liver*, 1st ed.; Patel, V.B., Rajendram, R., Preedy, V.R., Eds.; Academic Press: London, UK, 2018; pp. 15–20, ISBN 9780128039519.

99. Younossi, Z.; Anstee, Q.M.; Marietti, M.; Hardy, T.; Henry, L.; Eslam, M.; Bugianesi, E. Global burden of NAFLD and NASH: Trends, predictions, risk factors and prevention. *Nat. Rev. Gastroenterol. Hepatol.* **2018**, *15*, 11–20. [CrossRef]

100. Calzadilla Bertot, L.; Adams, L.A. The natural course of non-alcoholic fatty liver disease. *Int. J. Mol. Sci.* **2016**, *17*, 774. [CrossRef]

101. Mencin, A.A.; Lavine, J.E. Nonalcoholic fatty liver disease in children. *Curr. Opin. Clin. Nutr. Metab. Care* **2011**, *14*, 151–157. [CrossRef]

102. Sayiner, M.; Koenig, A.; Henry, L.; Younossi, Z.M. Epidemiology of nonalcoholic fatty liver disease and nonalcoholic steatohepatitis in the United States and the rest of the world. *Clin. Liver Dis.* **2016**, *20*, 205–214. [CrossRef]

103. Liu, W.; Baker, S.S.; Baker, R.D.; Zhu, L. Antioxidant mechanisms in nonalcoholic fatty liver disease. *Curr. Drug Targets* **2015**, *16*, 1301–1314. [CrossRef]

104. Rolo, A.P.; Teodoro, J.S.; Palmeira, C.M. Role of oxidative stress in the pathogenesis of nonalcoholic steatohepatitis. *Free Radic. Biol. Med.* **2012**, *52*, 59–69. [CrossRef]

105. Shiota, G.; Tsuchiya, H. Pathophysiology of NASH: Insulin resistance, free fatty acids and oxidative stress. *J. Clin. Biochem. Nutr.* **2006**, *38*, 127–132. [CrossRef]

106. Ota, T.; Takamura, T.; Kurita, S.; Matsuzawa, N.; Kita, Y.; Uno, M.; Nakanuma, Y. Insulin resistance accelerates a dietary rat model of nonalcoholic steatohepatitis. *Gastroenterology* **2007**, *132*, 282–293. [CrossRef]

107. Tilg, H.; Moschen, A.R. Evolution of inflammation in nonalcoholic fatty liver disease: The multiple parallel hits hypothesis. *Hepatology* **2010**, *52*, 1836–1846. [CrossRef]

108. Cusi, K. Role of obesity and lipotoxicity in the development of nonalcoholic steatohepatitis: Pathophysiology and clinical implications. *Gastroenterology* **2012**, *142*, 711–725. [CrossRef]

109. Karadeniz, G.; Acikgoz, S.; Tekin, I.O.; Tascýlar, O.; Gun, B.D.; Cömert, M. Oxidized low-density-lipoprotein accumulation is associated with liver fibrosis in experimental cholestasis. *Clinics* **2008**, *63*, 531–540. [CrossRef]

110. Fon Tacer, K.; Rozman, D. Nonalcoholic Fatty liver disease: Focus on lipoprotein and lipid deregulation. *J. Lipids* **2011**, *2011*, 783976. [CrossRef]

111. Werman, M.J.; Ben-Amotz, A.; Mokady, S. Availability and antiperoxidative effects of beta-carotene from Dunaliella bardawil in alcohol-drinking rats. *J. Nutr. Biochem.* **1999**, *10*, 449–454. [CrossRef]

112. Chen, H.; Tappel, A. Protection by multiple antioxidants against lipid peroxidation in rat liver homogenate. *Lipids* **1996**, *31*, 47–50. [CrossRef]

113. Whittaker, P.; Wamer, W.G.; Chanderbhan, R.F.; Dunkel, V.C. Effects of alpha-tocopherol and beta-carotene on hepatic lipid peroxidation and blood lipids in rats with dietary iron overload. *Nutr. Cancer* **1996**, *25*, 119–128. [CrossRef]

114. Senoo, H.; Yoshikawa, K.; Morii, M.; Miura, M.; Imai, K.; Mezaki, Y. Hepatic stellate cell (vitamin A-storing cell) and its relative–past, present and future. *Cell Biol. Int.* **2010**, *34*, 1247–1272. [CrossRef]

115. Xiao, M.L.; Chen, G.D.; Zeng, F.F.; Qiu, R.; Shi, W.Q.; Lin, J.S.; Chen, Y.M. Higher serum carotenoids associated with improvement of non-alcoholic fatty liver disease in adults: A prospective study. *Eur. J. Nutr.* **2019**, *58*, 721–730. [CrossRef]

116. Burri, B.J. Lycopene and human health. In *Phytochemicals in Nutrition and Health*, 1st ed.; Meskin, M.S., Bidlack, W.R., Davies, A.J., Omaye, S.T., Eds.; CRC Press: Boca Raton, FL, USA, 2002; pp. 157–172, ISBN 9781587160837.

117. Sarni, R.O.; Suano de Souza, F.I.; Ramalho, R.A.; Schoeps Dde, O.; Kochi, C.; Catherino, P.; Colugnati, F.B. Serum retinol and total carotene concentrations in obese pre-school children. *Med. Sci. Monit.* **2005**, *11*, CR510–CR514.

118. Seif El-Din, S.H.; El-Lakkany, N.M.; El-Naggar, A.A.; Hammam, O.A.; Abd El-Latif, H.A.; Ain-Shoka, A.A.; Ebeid, F.A. Effects of rosuvastatin and/or β-carotene on non-alcoholic fatty liver in rats. *Res. Pharm. Sci.* **2015**, *10*, 275–287.

119. Baybutt, R.C.; Molteni, A. Dietary beta-carotene protects lung and liver parenchyma of rats treated with monocrotaline. *Toxicology* **1999**, *137*, 69–80. [CrossRef]

120. Patel, V.; Sail, S. β-carotene protects the physiological antioxidants against aflatoxin-B1 induced carcinogenesis in albino rats. *Pak. J. Biol. Sci.* **2006**, *9*, 1104–1111.

121. Kheir Eldin, A.A.; Motawi, T.M.; Sadik, N.A. Effect of some natural antioxidants on aflatoxin B1-induced hepatic toxicity. *EXCLI J.* **2008**, *7*, 119–131.

122. Harari, A.; Harats, D.; Marko, D.; Cohen, H.; Barshack, I.; Kamari, Y.; Shaish, A. A 9-cis beta-carotene-enriched diet inhibits atherogenesis and fatty liver formation in LDL receptor knockout mice. *J. Nutr.* **2008**, *138*, 1923–1930. [CrossRef]

123. Ozturk, F.; Gul, M.; Ates, B.; Ozturk, I.C.; Cetin, A.; Vardi, N.; Yilmaz, I. Protective effect of apricot (*Prunus armeniaca* L.) on hepatic steatosis and damage induced by carbon tetrachloride in Wistar rats. *Br. J. Nutr.* **2009**, *102*, 1767–1775. [CrossRef]

124. Liu, Q.; Bengmark, S.; Qu, S. Nutrigenomics therapy of hepatisis C virus induced-hepatosteatosis. *BMC Gastroenterol.* **2010**, *10*, 49. [CrossRef]

125. Yadav, D.; Hertan, H.I.; Schweitzer, P.; Norkus, E.P.; Pitchumoni, C.S. Serum and liver micronutrient antioxidants and serum oxidative stress in patients with chronic hepatitis C. *Am. J. Gastroenterol.* **2002**, *97*, 2634–2639. [CrossRef]

126. Tainaka, T.; Shimada, Y.; Kuroyanagi, J.; Zang, L.; Oka, T.; Nishimura, Y.; Tanaka, T. Transcriptome analysis of anti-fatty liver action by Campari tomato using a zebrafish diet-induced obesity model. *Nutr. Metab.* **2011**, *8*, 88. [CrossRef]

127. Xiao, J.; Liong, E.C.; Ching, Y.P.; Chang, R.C.C.; Fung, M.L.; Xu, A.M.; Tipoe, G.L. Lycium barbarum polysaccharides protect rat liver from non-alcoholic steatohepatitis-induced injury. *Nutr. Diabetes* **2013**, *3*, e81. [CrossRef]

128. Villaça Chaves, G.; Pereira, S.E.; Saboya, C.J.; Ramalho, A. Non-alcoholic fatty liver disease and its relationship with the nutritional status of vitamin A in individuals with class III obesity. *Obes. Surg.* **2008**, *18*, 378–385. [CrossRef]

129. Lan, Q.Y.; Zhang, Y.J.; Liao, G.C.; Zhou, R.F.; Zhou, Z.G.; Chen, Y.M.; Zhu, H.L. The Association between Dietary Vitamin A and Carotenes and the Risk of Primary Liver Cancer: A Case–Control Study. *Nutrients* **2016**, *8*, 624. [CrossRef]

130. Wang, L.; Ding, C.; Zeng, F.; Zhu, H. Low Levels of Serum β-Carotene and β-Carotene/Retinol Ratio Are Associated with Histological Severity in Nonalcoholic Fatty Liver Disease Patients. *Ann. Nutr. Metab.* **2019**, *74*, 156–164. [CrossRef]

131. Britton, G. Structure and properties of carotenoids in relation to function. *FASEB J.* **1995**, *9*, 1551–1558. [CrossRef]

132. Heber, D.; Lu, Q.Y. Overview of mechanisms of action of lycopene. *Exp. Biol. Med.* **2002**, *227*, 920–923. [CrossRef]

133. Stahl, W.; Heinrich, U.; Aust, O.; Tronnier, H.; Sies, H. Lycopene-rich products and dietary photoprotection. *Photochem. Photobiol. Sci.* **2006**, *5*, 238–242. [CrossRef]

134. Shivashangari, K.S.; Ravikumar, V.; Vinodhkumar, R.; Sheriff, S.A.; Devaki, T. Hepatoprotective potential of lycopene on D-galactosamine/lipopolysaccharide induced hepatitis in rats. *Pharmacologyonline* **2006**, *2*, 151–170.

135. Sheriff, S.A.; Devaki, T. Lycopene stabilizes lipoprotein levels during D-galactosamine/lipopolysaccharide induced hepatitis in experimental rats. *Asian Pac. J. Trop. Biomed.* **2012**, *2*, 975–980. [CrossRef]

136. Wang, Y.; Ausman, L.M.; Greenberg, A.S.; Russell, R.M.; Wang, X.D. Dietary lycopene and tomato extract supplementations inhibit nonalcoholic steatohepatitis-promoted hepatocarcinogenesis in rats. *Int. J. Cancer* **2010**, *126*, 1788–1796. [CrossRef]

137. Ahn, J.; Lee, H.; Jung, C.H.; Ha, T. Lycopene inhibits hepatic steatosis via microRNA-21-induced downregulation of fatty acid-binding protein 7 in mice fed a high-fat diet. *Mol. Nutr. Food Res.* **2012**, *56*, 1665–1674. [CrossRef]

138. Bahcecioglu, I.H.; Kuzu, N.; Metin, K.; Ozercan, I.H.; Ustündag, B.; Sahin, K.; Kucuk, O. Lycopene prevents development of steatohepatitis in experimental nonalcoholic steatohepatitis model induced by high-fat diet. *Vet. Med. Int.* **2010**, *2010*, 262179. [CrossRef]

139. Bernal, C.; Martín-Pozuelo, G.; Lozano, A.B.; Sevilla, A.; García-Alonso, J.; Canovas, M.; Periago, M.J. Lipid biomarkers and metabolic effects of lycopene from tomato juice on liver of rats with induced hepatic steatosis. *J. Nutr. Biochem.* **2013**, *24*, 1870–1881. [CrossRef]

140. Ip, B.; Wang, X.D. Non-alcoholic steatohepatitis and hepatocellular carcinoma: Implications for lycopene intervention. *Nutrients* **2013**, *6*, 124–162. [CrossRef]

141. Piña-Zentella, R.M.; Rosado, J.L.; Gallegos-Corona, M.A.; Madrigal-Pérez, L.A.; García, O.P.; Ramos-Gomez, M. Lycopene improves diet-mediated recuperation in rat model of nonalcoholic fatty liver disease. *J. Med. Food* **2016**, *19*, 607–614. [CrossRef]

142. Kujawska, M.; Ewertowska, M.; Adamska, T.; Sadowski, C.; Ignatowicz, E.; Jodynis-Liebert, J. Antioxidant effect of lycopene-enriched tomato paste on *N*-nitrosodiethylamine-induced oxidative stress in rats. *J. Physiol. Biochem.* **2014**, *70*, 981–990. [CrossRef]

143. Ip, B.C.; Liu, C.; Ausman, L.M.; Von Lintig, J.; Wang, X.D. Lycopene attenuated hepatic tumorigenesis via differential mechanisms depending on carotenoid cleavage enzyme in mice. *Cancer Prev. Res.* **2014**, *7*, 1219–1227. [CrossRef]

144. Navarro-González, I.; García-Alonso, J.; Periago, M.J. Bioactive compounds of tomato: Cancer chemopreventive effects and influence on the transcriptome in hepatocytes. *J. Funct. Foods* **2018**, *42*, 271–280. [CrossRef]

145. Martín-Pozuelo, G.; Navarro-González, I.; González-Barrio, R.; Santaella, M.; García-Alonso, J.; Hidalgo, N.; Periago, M.J. The effect of tomato juice supplementation on biomarkers and gene expression related to lipid metabolism in rats with induced hepatic steatosis. *Eur. J. Nutr.* **2015**, *54*, 933–944. [CrossRef]

146. Bandeira, A.C.B.; Da Silva, T.P.; De Araujo, G.R.; Araujo, C.M.; Da Silva, R.C.; Lima, W.G.; Costa, D.C. Lycopene inhibits reactive oxygen species production in SK-Hep-1 cells and attenuates acetaminophen-induced liver injury in C57BL/6 mice. *Chem. Biol. Interact.* **2017**, *263*, 7–17. [CrossRef]

147. Bandeira, A.C.B.; Da Silva, R.C.; Júnior, J.V.R.; Figueiredo, V.P.; Talvani, A.; Cangussú, S.D.; Costa, D.C. Lycopene pretreatment improves hepatotoxicity induced by acetaminophen in C57BL/6 mice. *Bioorg. Med. Chem.* **2017**, *25*, 1057–1065. [CrossRef]

148. Yefsah-Idres, A.; Benazzoug, Y.; Otman, A.; Latour, A.; Middendorp, S.; Janel, N. Hepatoprotective effects of lycopene on liver enzymes involved in methionine and xenobiotic metabolism in hyperhomocysteinemic rats. *Food Funct.* **2016**, *7*, 2862–2869. [CrossRef]

149. Gupta, P.; Bhatia, N.; Bansal, M.P.; Koul, A. Lycopene modulates cellular proliferation, glycolysis and hepatic ultrastructure during hepatocellular carcinoma. *World J. Hepatol.* **2016**, *8*, 1222–1233. [CrossRef]

150. Xu, F.; Yu, K.; Yu, H.; Wang, P.; Song, M.; Xiu, C.; Li, Y. Lycopene relieves AFB1-induced liver injury through enhancing hepatic antioxidation and detoxification potential with Nrf2 activation. *J. Funct. Foods* **2017**, *39*, 215–224. [CrossRef]

151. Elvira-Torales, L.I.; Navarro-González, I.; González-Barrio, R.; Martín-Pozuelo, G.; Doménech, G.; Seva, J.; Periago-Castón, M. Tomato juice supplementation influences the gene expression related to steatosis in rats. *Nutrients* **2018**, *10*, 1215. [CrossRef]

152. Landrum, J.T.; Bone, R.A. Lutein, zeaxanthin, and the macular pigment. *Arch. Biochem. Biophys.* **2001**, *385*, 28–40. [CrossRef]

153. Pang, R.; Tao, J.Y.; Zhang, S.L.; Zhao, L.; Yue, X.; Wang, Y.F.; Wu, J.G. In vitro antiviral activity of lutein against hepatitis B virus. *Phytother. Res.* **2010**, *24*, 1627–1630. [CrossRef]

154. Kim, J.E.; Clark, R.M.; Park, Y.; Lee, J.; Fernandez, M.L. Lutein decreases oxidative stress and inflammation in liver and eyes of guinea pigs fed a hypercholesterolemic diet. *Nutr. Res. Pract.* **2012**, *6*, 113–119. [CrossRef]

155. Sindhu, E.R.; Firdous, A.P.; Preethi, K.C.; Kuttan, R. Carotenoid lutein protects rats from paracetamol-, carbon tetrachloride-and ethanol-induced hepatic damage. *J. Pharm. Pharmacol.* **2013**, *62*, 1054–1060. [CrossRef]

156. Qiu, X.; Gao, D.H.; Xiang, X.; Xiong, Y.F.; Zhu, T.S.; Liu, L.G.; Hao, L.P. Ameliorative effects of lutein on non-alcoholic fatty liver disease in rats. *World J. Gastroenterol.* **2015**, *21*, 8061–8072. [CrossRef]

157. Murillo, A.G.; Aguilar, D.; Norris, G.H.; DiMarco, D.M.; Missimer, A.; Hu, S.; Fernandez, M.L. Compared with powdered lutein, a lutein nanoemulsion increases plasma and liver lutein, protects against hepatic steatosis, and affects lipoprotein metabolism in guinea pigs. *J. Nutr.* **2016**, *146*, 1961–1969. [CrossRef]

158. Haegele, A.D.; Gillette, C.; O'Neill, C.; Wolfe, P.; Heimendinger, J.; Sedlacek, S.; Thompson, H.J. Plasma xanthophyll carotenoids correlate inversely with indices of oxidative DNA damage and lipid peroxidation. *Cancer Epidemiol. Biomark. Prev.* **2000**, *9*, 421–425.

159. Katsuura, S.; Imamura, T.; Bando, N.; Yamanishi, R. Beta-Carotene and beta-cryptoxanthin but not lutein evoke redox and immune changes in RAW264 murine macrophages. *Mol. Nutr. Food Res.* **2009**, *53*, 1396–1405. [CrossRef]

160. Takayanagi, K. Prevention of adiposity by the oral administration of β-cryptoxanthin. *Front. Neurol.* **2011**, *2*, 67. [CrossRef]

161. Kobori, M.; Ni, Y.; Takahashi, Y.; Watanabe, N.; Sugiura, M.; Ogawa, K.; Ota, T. β-Cryptoxanthin alleviates diet-induced nonalcoholic steatohepatitis by suppressing inflammatory gene expression in mice. *PLoS ONE* **2014**, *9*, e98294. [CrossRef]

162. Ni, Y.; Nagashimada, M.; Zhan, L.; Nagata, N.; Kobori, M.; Sugiura, M.; Ota, T. Prevention and reversal of lipotoxicity-induced hepatic insulin resistance and steatohepatitis in mice by an antioxidant carotenoid, β-cryptoxanthin. *Endocrinology* **2015**, *156*, 987–999. [CrossRef]

163. Murakoshi, M.; Nishino, H.; Satomi, Y.; Takayasu, J.; Hasegawa, T.; Tokuda, H.; Iwasaki, R. Potent preventive action of α-carotene against carcinogenesis: Spontaneous liver carcinogénesis and promoting stage of lung and skin carcinogenesis in mice are suppressed more effectively by α-carotene than by β-carotene. *Cancer Res.* **1992**, *52*, 6583–6587.

164. Chamberlain, S.M.; Hall, J.D.; Patel, J.; Lee, J.R.; Marcus, D.M.; Sridhar, S.; Bartoli, M. Protective effects of the carotenoid zeaxanthin in experimental nonalcoholic steatohepatitis. *Dig. Dis. Sci.* **2009**, *54*, 1460–1464. [CrossRef]

165. Cao, Y.; Wang, C.; Liu, J.; Liu, Z.M.; Ling, W.H.; Chen, Y.M. Greater serum carotenoid levels associated with lower prevalence of nonalcoholic fatty liver disease in Chinese adults. *Sci. Rep.* **2015**, *5*, 12951. [CrossRef]

antioxidants

MDPI

Article

Lycopene Modulates Pathophysiological Processes of Non-Alcoholic Fatty Liver Disease in Obese Rats

Mariane Róvero Costa [1,*], Jéssica Leite Garcia [1], Carol Cristina Vágula de Almeida Silva [1], Artur Junio Togneri Ferron [1], Fabiane Valentini Francisqueti-Ferron [1], Fabiana Kurokawa Hasimoto [1], Cristina Schmitt Gregolin [1], Dijon Henrique Salomé de Campos [1], Cleverton Roberto de Andrade [2], Ana Lúcia dos Anjos Ferreira [1], Camila Renata Corrêa [1] and Fernando Moreto [1]

[1] Medical School, São Paulo State University (Unesp), Botucatu 18618-687, Brazil
[2] School of Dentistry, São Paulo State University (Unesp), Araraquara 14800-901, Brazil
* Correspondence: marianerovero@gmail.com; Tel.: +55-14-3880-1722

Received: 27 June 2019; Accepted: 1 August 2019; Published: 5 August 2019

Abstract: *Background*: The higher consumption of fat and sugar are associated with obesity development and its related diseases such as non-alcoholic fatty liver disease (NAFLD). Lycopene is an antioxidant whose protective potential on fatty liver degeneration has been investigated. The aim of this study was to present the therapeutic effects of lycopene on NAFLD related to the obesity induced by a hypercaloric diet. *Methods*: Wistar rats were distributed in two groups: Control (Co, $n = 12$) and hypercaloric (Ob, $n = 12$). After 20 weeks, the animals were redistributed into the control group (Co, $n = 6$), control group supplemented with lycopene (Co+Ly, $n = 6$), obese group (Ob, $n = 6$), and obese group supplemented with lycopene (Ob+Ly, $n = 6$). Ob groups also received water + sucrose (25%). Animals received lycopene solution (10 mg/kg/day) or vehicle (corn oil) via gavage for 10 weeks. *Results*: Animals which consumed the hypercaloric diet had higher adiposity index, increased fasting blood glucose, hepatic and blood triglycerides, and also presented in the liver macro and microvesicular steatosis, besides elevated levels of tumor necrosis factor-α (TNF-α). Lycopene has shown therapeutic effects on blood and hepatic lipids, increased high-density lipoprotein cholesterol (HDL), mitigated TNF-α, and malondialdehyde (MDA) and further improved the hepatic antioxidant capacity. *Conclusion*: Lycopene shows therapeutic potential to NAFLD.

Keywords: metabolic syndrome; NAFLD; hypercaloric diet; obesity

1. Introduction

Obesity is a chronic disease with high prevalence worldwide, which represents an overwhelming problem for public health [1,2]. Recent estimates indicate that more than 1.9 billion adults are overweight and of these, more than 650 million are obese [3]. Although weight control is a consequence of a complex interaction between genetic, metabolic, psychological, and social factors, the obese phenotype is broadly influenced by behavioral aspects [4,5]. The energy imbalance resulting from the hypercaloric foods consumption and sedentary lifestyle corroborates for such an outcome [6]. The higher intake of processed foods containing saturated fat, added sugar, and high energy density led to an increase in the caloric value of the diet [7]. Taken together these factors favor the development of obesity and its related diseases such as nonalcoholic fatty liver disease (NAFLD) [8,9].

NAFLD comprises a spectrum of diseases ranging from simple steatosis to biochemical and functional abnormalities in the liver, which may result in an inflammatory process known as nonalcoholic steatohepatitis (NASH). If this problem is not solved, it may progress to hepatic cirrhosis and hepatocellular carcinoma [9]. It is hypothesized that these liver changes are the result of a two-step process, named the "two hit" theory. The first hit results from the imbalance between hepatic lipid influx and hepatic lipid clearance. The second hit involves the inflammatory and oxidative process [10,11]. Another hypothesis attributed to the NAFLD genesis is the "multiple parallel-hit", which comprises several parallel hits including insulin resistance, inflammation, oxidative stress, genetic and epigenetic mechanisms, environmental elements, and microbiota changes [11]. Except for some divergences, these hypotheses present common mechanisms, including oxidative stress and inflammation, which play a central role in the genesis of NAFLD.

Dietary factors are also important to liver fat deposition [10]. The excess of nutrients received by the liver are converted into triglycerides, which can be oxidized or exported to extra-hepatic organs [12,13]. In the presence of huge amounts of the substrate, the mitochondria increase the production of reactive oxygen species (ROS), impairing its function [14]. ROS are capable of activating inflammatory pathways and the inflammatory cells increase the ROS production, which leads to tissue damage [15]. Moreover, components of the diet itself can directly lead to the activation of inflammatory pathways, such as saturated fatty acids [16], or also lead to intestinal dysbiosis and consequent infiltration of lipolysaccharides into the bloodstream, contributing to the inflammatory process [17]. Considering this overview and the pathological consequences of NAFLD, it is important to seek effective alternatives to treat or prevent hepatic accumulation of lipids and its unfolding. A wide range of drugs are available for the treatment of NAFLD in order to avoid its progress, including the use of antioxidants, such as lycopene [18].

Lycopene is a carotenoid without provitamin A activity, which gives red colors to vegetables such as tomatoes, red grapefruit, watermelon, and apricots [19]. This carotenoid is a powerful antioxidant with anti-inflammatory activity [20], which has been shown to modulate important metabolic processes in the body, attenuating complications of obesity and restoring liver function. The lycopene is a liposoluble molecule, which depends on dietary lipids to be absorbed. The liver is one of the major organs that stores lycopene, so in this point of view, the investigation of lycopene effects on NAFLD is valuable [19]. Thus, the aim of this study is to verify lycopene's therapeutic potential on the pathophysiological processes of NAFLD in obese animals fed a hypercaloric diet.

2. Materials and Methods

2.1. Animals and Experimental Protocol

This studied was approved by the Ethics Committee on the Use of Animals (CEUA protocol number 1266/2018) of Botucatu Medical School, Sao Paulo State University. The experiment was performed in accordance with the National Institute of Health's Guide to the Care and Use of Laboratory Animals [21]. Young male Wistar rats (8 weeks old) were kept in a controlled environment of temperature (22 ± 3 °C), luminosity (12 h light-dark cycle), and humidity (relative humidity of 60 ± 5%). Initially, the animals were distributed in two experimental groups fed different diets for induction of obesity: Control diet ($n = 12$) and hypercaloric diet ($n = 12$). At the 20th week, after confirmation of difference in body weight and plasma triglycerides, the animals were redistributed into 4 groups for the lycopene supplementation study: A control group (Co, $n = 6$), a control group supplemented with lycopene (Co+Ly, $n = 6$), an obese group (Ob, $n = 6$), and an obese group supplemented with lycopene (Ob+Ly, $n = 6$). The feed and water were ad libitum. The groups received lycopene solution (10 mg/kg/day) or vehicle (corn oil) via gavage for 10 weeks.

2.2. Diets

The diets used in this study were previously described by Francisqueti et al., (2017) [22]. Briefly, both diets were nutritionally balanced for micronutrients but different for macronutrients. The control diet was composed of soybean meal, sorghum, soybean peel, dextrin, soybean oil, vitamins, and minerals. The hypercaloric diet was designed to mimic the development of obesity associated with western dietary habits, being composed of soybean meal, sorghum, soybean peel, dextrin, sucrose, fructose, lard, vitamins and minerals, and addition of 25% sucrose in the drinking water.

2.3. Lycopene Supplementation

The lycopene solution used in this study was the tomato oleoresin (Lyc-O-Mato® 6% dewaxed) obtained from LycoRed Natural Products Industries, Beersheba, Israel. The solution was mixed with corn oil, checked for desired concentration absorbance (450 nm) and stored in a dark environment to ensure solution stability. The animals received lycopene solution (10 mg/kg/day) via gavage for 10 weeks [23,24]. The non-lycopene supplemented groups received corn oil vehicle (1 mL) via gavage.

2.4. Euthanasia

At the end of the 30-week period, after 8 h of fasting the animals were anesthetized (thiopental 120 mg/kg/i.p.) and euthanized by decapitation after the absence of foot reflex to obtain blood, liver, and adipose tissues (visceral, retroperitoneal and epididymal). Blood samples were collected in Falcon® tubes (Kasvi, São José dos Pinhais, Paraná, Brazil) containing anticoagulant ethylenediamine tetraacetic acid (EDTA) and centrifuged at 2000× g and 4 °C for 10 min (Eppendorf® Centrifuge 5804-R, Hamburg, Germany) and the plasma was stored in a freezer at −80 °C. The liver was fractionated in aliquots, the left hepatic lobe was stored in formaldehyde for histological analysis, and the remaining aliquots were stored in sterile cryotubes in a freezer at −80 °C. The fat deposits were dissected for weighing.

2.5. Nutritional and Obesity Characterization

The nutritional characterization was evaluated considering the chow intake, water intake, and caloric intake (diet + water). For that, water and chow were measured twice a week. The caloric intake was calculated by multiplying the daily chow intake by the energy value of each diet. The presence of obesity was established based on weight gain and adiposity index. For that, the body weight was measured weekly. The weight gain was calculated by subtracting the initial weight from the final weight of the animals [weight gain (g) = final weight (g) − initial weight (g)]. The adiposity index represents the ratio of the sum of the epididymal, visceral, and retroperitoneal fat deposits by the final weight multiplied by 100 [adiposity index (%) = (epididymal (g) + visceral (g) + retroperitoneal (g))/final weight (g) × 100].

2.6. Lycopene Bioavailability Evaluation

The presence of lycopene was determined in plasma and hepatic tissue homogenate. To extract the carotenoids, samples were incubated with internal standard (equinenone), chloroform/methanol $CHCl_3/CH_3OH$ (3 mL, 2:1, v/v) and 500 mL of saline 8.5 g/L. Then the samples were centrifuged at 2000× g for 10 min and the supernatant was collected and hexane was added. The chloroform and hexane layers were evaporated under nitrogen and the residue was resuspended in 150 mL of ethanol and sonicated for 30 s. 50 µL of this aliquot was injected into the HPLC. The HPLC system was a Waters Alliance 2695 (Waters, Wilmington, MA, USA) and consisted of pump and chromatography bound to a 2996 programmable photodiode array detector, a C30 carotenoid column (3 mm, 150 × 3 × 4.6 mm, YMC, Wilmington, MA, USA) and Empower 3 chromatography data software (Milford, MA, USA). The HPLC system programmable photodiode array detector was set at 450 nm for carotenoids. The mobile phase consisted of ethanol/methanol/methyl-tert-butyl ether/water (83:15:2, $v/v/v$, 15 g/L with ammonium acetate in water, solvent A) and methanol/methyl-tert-butyl ether/water (8:90:2, $v/v/v$, 10 g/L with ammonium acetate in water, solvent B). The gradient procedure, at a flow rate of 1 mL/min (16 °C), was as follows: (1) 100% solvent A was used for 2 min followed by a 6 min linear gradient to 70% solvent A; (2) a 3 min hold followed by a 10 min linear gradient to 45% solvent A; (3) a 2 min hold, then a 10 min linear gradient to 5% solvent A; (4) a 4 min hold, then a 2 min linear gradient back to 100% solvent A. For the quantification of the chromatograms, a comparison was made between the area ratio of the substance and area of the internal standard obtained in the analysis [25].

2.7. Clinical Biochemistry

The glucose determination was performed by glycosimeter (Accu-Chek Performa; Roche Diagnostics Indianopolis, IN, USA). The concentrations of urea, creatinine, uric acid, total proteins, albumin, aspartate aminotransferase (AST), alanine aminotransferase (ALT), triglycerides, total cholesterol and its fractions, high-density lipoprotein cholesterol (HDL), and non-HDL cholesterol, were performed by an enzymatic colorimetric method with commercial kits (BioClin®, Belo Horizonte, MG, Brazil) in an automatic enzymatic analyzer system (Chemistry Analyzer BS-200, Mindray Medical International Limited, Shenzhen, China).

2.8. Hepatic Tissue Analysis

2.8.1. Histology

Hepatic tissue was stored in 4% paraformaldehyde with 0.1 M phosphate buffer (pH 7.4) during the first 24 h. After, tissue was transferred to 70% ethyl alcohol until paraffin waxing. Histological sections obtained from the paraffin block were laid on slides and stained with hematoxylin and eosin (H&E). Macrovesicular steatosis and microvesicular steatosis were both separately scored and the severity was graded based on the percentage of the total area affected. The scores were: 0 (<5), 1 (≥5–33%), 2 (≥33–66%) and 3 (≥66%). The difference between macrovesicular and microvesicular steatosis was defined by whether the vacuoles displaced the nucleus to the side (macrovesicular) or not (microvesicular) [26]. Ten fields were analyzed per slide per animal. For the statistical analysis, it was considered the sum of the scores obtained in the 10 analyzed fields.

2.8.2. Total Proteins

The tissue samples were homogenized in phosphate-buffered saline (PBS) at a ratio of 1:10 (sample:PBS). Total protein concentrations in hepatic tissue were measured in the homogenate by a colorimetric method using the biuret reagent (BioClin, Quibasa Química Básica Ltd.a., Belo Horizonte, Minas Gerais, Brazil) and were verified in an automatic enzymatic analyzer system (Chemistry Analyzer BS-200, Mindray Medical International Ltd., Shenzhen, China). The results were used to correct the parameters of the inflammation and redox state.

2.8.3. Inflammatory Biomarkers

The hepatic concentrations of interleukin-6 (IL-6) and tumor necrosis factor-α (TNF-α) were quantified in the tissue homogenate by enzyme-linked immunosorbent assay (ELISA) using specific commercial kits (R&D Systems®, Minneapolis, MN, USA) and a micro-plate reader (Spectra MAX 190, Molecular Devices, Sunny Valley, CA, USA). The results were corrected for the total protein present in the tissue and expressed in picogram per milligram of protein (pg/mg protein). The tissue samples were homogenized in phosphate-buffered saline (PBS) at the ratio of 1:10 (sample:PBS).

2.8.4. Redox State

Oxidative Damage to Lipids

The malondialdehyde (MDA) concentration was obtained by high performance liquid chromatography with fluorometric detection (HPLC, system LC10A, Shimadzu, Japan). The hepatic tissue homogenate (100 μL) was incubated with thiobarbituric acid (TBA 42 mmol/L, 200 μL) and 1% ortho-phosphoric acid (700 μL) at 100 °C for 60 min to form the fluorescent adduct TBA-MDA. Deproteinization was performed with 2 mmol/L sodium hydroxide and methanol (NaOH:MetOH; 1:1) submitted to centrifugation (2000× g, 5 min). After 50 μL of the supernatant was filtered and injected into an octadecylsilic column (ODS-2, 150 × 4.6 mm, 5 μm; Spherisarb®, Waters). The isocratic mobile phase comprised of the phosphate buffer (10 mmol/L, pH:6.8) and HPLC grade methanol (60:40 v/v), it was run through the system at a flow rate of 0.5 mL/min. The calibration curve was obtained with the preparation of tetra-ethoxy propane solutions (TEP). Fluorimetric detection was performed at 527 nm of excitation and 551 nm of emission. The results were expressed as nanomol per milligram protein (nmol/mg protein) [27].

Antioxidant Enzyme Activity

In these analyzes, the activity of superoxide dismutase (SOD) and catalase (CAT) was evaluated. Therefore, 100mg liver was homogenized (1:20 *v/v*) in KH_2PO_4 (10 mmol/L)/KCl (120 mmol/L), pH 7.4, and centrifuged at 2.000× *g* for 20 min. SOD activity was measured based on the inhibition of a superoxide radical reaction with pyrogallol, and the absorbance values were measured at 420 nm [28]. The values were expressed as units per milligram of protein (U/mg protein). CAT activity was evaluated by following the decrease in the levels of hydrogen peroxide, absorbance was measured at 240 nm [29]. The results were expressed as picomol per milligram of protein (pmol/mg protein).

Hepatic Antioxidant Capacity

The hepatic tissue antioxidant capacity was determined by the total antioxidant performance test (TAP) using a VICTOR X2 reader (Perkin Elmer-Boston, MA, USA) [30]. This assay utilized 100 μL of homogenate, which was incubated with the fluorescent indicator BODIPY (4,4-difluoro-4-bora-3a, 4a-diaza-s-indacene) for 10 min at 37 °C. Then the free radical generator 2,2′Azobis(2-amidino-propane)-dihydrochloride (AAPH) was added to the samples. The prepared samples were applied in triplicates, 200 μL in each well on specific plates. Phosphatidylcholine (PC) was used as a hydrophilic matrix reference. The fluorescence reader (Wallac Vitor 2X®, Perkin-Elmer, Boston, MA, USA) was programmed to perform readings every 5 min for 3 h and 30 min. The analyses were performed in triplicate, and the results represented the percent protection.

2.9. Statistical Analysis

Data were presented as means ± standard deviation (SD) or median (interquartile range). Differences among the groups were determined by two-way ANOVA with Holm-Sidak post-hoc test or by Kruskall Wallis test with Tukey post-hoc test. The comparison between the 2 groups was performed by the *T*-test. These statistical analyses were performed using the software Sigma Stat for Windows Version 3.5 (Systat Software Inc., San Jose, CA, USA). For the histological parameters, the Poasson distribution or the binomial distribution followed by the post-hoc test Wald multi-comparison were used. These statistical analyses were performed by an experienced statistician using the software Statistical Analysis System (SAS) 9.4 (SAS Institute Inc., Campus Drive Cary, NC, USA). A *p* value of ≤0.05 was considered statistically significant.

3. Results

3.1. Lycopene Does Not Influence Nutritional and Adiposity Markers

Both Ob and Ob+Ly groups presented lower chow intake. There was no difference between the groups in water and caloric intake. There was no difference between the initial weights. The obese groups (Ob and Ob+Ly) presented a higher final weight and weight gain (Table 1).

The determination of the presence of obesity is shown in Figure 1. Similar to weight gain, the adiposity index was higher in the groups Ob and Ob+Ly.

Table 1. Intake and weight characteristics.

Parameters	Co	Co+Ly	Ob	Ob+Ly
Chow intake (g/d)	24 ± 0.8 [a]	24.4 ± 0.4 [a]	12.8 ± 0.6 [b]	11.8 ± 0.4 [b]
Water intake (mL/d)	24.1 ± 0.6 [a]	27 ± 1.4 [a]	23.9 ± 0.6 [a]	23.7 ± 0.4 [a]
Caloric intake (kcal/d)	89.9 ± 2.9 [a]	88.9 ± 0.9 [a]	95 ± 3.9 [a]	94.2 ± 1.8 [a]
Initial Weight (g)	192 ± 28.4 [a]	209 ± 14.3 [a]	212 ± 21.4 [a]	198 ± 11.6 [a]
Final Weight (g)	490 ± 43.7 [a]	470 ± 59.6 [a]	568.5 ± 75.2 [b]	552 ± 42.5 [b]
Weight Gain (g)	297 ± 33.7 [a]	260 ± 52.0 [a]	356 ± 72.9 [b]	354 ± 34.2 [b]

Co: Control group; Co+Ly: Control group supplemented with lycopene; Ob: Obese group; Ob+Ly: hypercaloric group supplemented with lycopene. Data parametric expressed in mean ± standard deviation. Comparison by two-way ANOVA with Holm Sidak post-hoc test. Different letters correspond to the significant statistical difference (p <0.05); n = 6 animals/group.

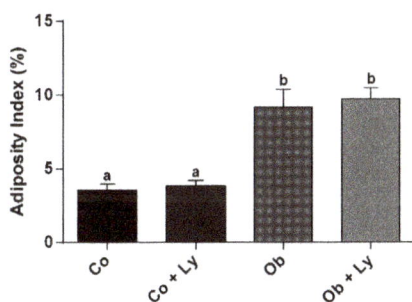

Figure 1. Determination of the presence of obesity: Adiposity index (%). Co: Control group; Co+Ly: control group supplemented with lycopene; Ob: Obese group; Ob+Ly: Hypercaloric group supplemented with lycopene. Data expressed in mean ± standard deviation. Comparison by two-way ANOVA with Holm-Sidak post-hoc test. Different letters correspond to the significant statistical difference (p <0.05); n = 6 animals/group.

3.2. Lycopene Is Available in Plasma and Liver of Supplemented Rats

The lycopene bioavailability is shown in Table 2. It is possible to verify the presence of lycopene in both groups, which were supplemented (Co+Ly and Ob+Ly).

Table 2. Lycopene concentration in hepatic tissue and plasma.

Parameters	Co	Co+Ly	Ob	Ob+Ly
Liver (µg/100 g tissue)	ND	47.32 ± 5.95 [a]	ND	25 ± 2.91 [b]
Plasma (µg/mL)	ND	3.18 ± 0.586 [a]	ND	5.22 ± 2.31 [a]

Co: Control group; Co+Ly: Control group supplemented with lycopene; Ob: Obese group; Ob+Ly: Hypercaloric group supplemented with lycopene. ND: Not detectable. Data parametric expressed in mean ± standard deviation. Comparison by t-test. Different letters correspond to the significant statistical difference (p < 0.05); n = 6 animals/group.

3.3. Lycopene Influences Plasma Lipid Markers

Table 3 presents the biochemical analysis related to renal and hepatic functions. There was no difference between the groups for all the parameters presented in Table 3.

Markers for lipid and glucose metabolism are exhibited in Table 4. The fasting blood glucose was higher in the obese groups (Ob and Ob+Ly) independent of lycopene supplementation. Plasma triglycerides were higher only in the Ob group. Similarly, the Ob group presented higher hepatic triglycerides levels, and the Ob+Ly group showed similar levels to the control groups (Co and Co+Ly). The total cholesterol was higher in the groups supplemented with lycopene. This finding could be influenced by increasing HDL cholesterol levels also observed in supplemented groups, mainly the Ob+Ly group. There was no difference between groups for non-HDL cholesterol (Table 4).

Table 3. Markers for clinical biochemistry.

Parameters	Co	Co+Ly	Ob	Ob+Ly
Urea (mg/dL)	54.2 ± 8.8 [a]	51.5 ± 3.9 [a]	48.0 ± 33.6 [a]	42.8 ± 20.9 [a]
Creatinine (mg/dL)	0.432 ± 0.034 [a]	0.433 ± 0.057 [a]	0.568 ± 0.291 [a]	0.504 ± 0.092 [a]
Uric Acid (mg/dL)	0.544 ± 0.103 [a]	0.693 ± 0.307 [a]	0.695 ± 0.122 [a]	0.800 ± 0.166 [a]
AST (U/L)	149 (100–230) [a]	123 (113–152) [a]	112 (87–213) [a]	141 (135–179) [a]
ALT (U/L)	44.5 (38.7–161) [a]	50.5 (35.5–106) [a]	51 (32.7–133) [a]	38.5 (32.5–95.7) [a]
Albumin (g/dL)	2.7 ± 0.1 [a]	2.6 ± 0.1 [a]	2.7 ± 0.1 [a]	2.7 ± 0.1 [a]
Total Proteins (g/dL)	5.6 ± 0.1 [a]	5.6 ± 0.3 [a]	5.9 ± 0.2 [a]	5.9 ± 0.3 [a]

Co: Control group; Co+Ly: Control group supplemented with lycopene; Ob: Obese group; Ob+Ly: Hypercaloric group supplemented with lycopene. AST: Aspartate aminotransferase; ALT: Alanine aminotransferase. Data parametric expressed in mean ± standard deviation. Comparison by two-way ANOVA with Holm Sidak *post-hoc* test. Data non-parametric expressed in median and interquartile range. Comparison by Kruskall Wallis test with Tukey post-hoc test. Different letters correspond to the significant statistical difference ($p < 0.05$); $n = 6$ animals/group.

Table 4. Markers for lipid and glucose metabolism.

Parameters	Co	Co+Ly	Ob	Ob+Ly
Fasting Blood Glucose (mg/dL)	74.2 ± 7.56 [a]	91.6 ± 16.6 [a]	102 ± 21.4 [b]	104 ± 7.50 [b]
Triglycerides (mg/dL)	62.0 ± 20.3 [a]	77.3 ± 32.5 [a]	113 ± 41.8 [b]	93.9 ± 12.4 [a]
Hepatic Triglycerides (mg/dL)	20.6 ± 4.29 [a]	20.6 ± 2.92 [a]	33.2 ± 7.9 [b]	29.3 ± 5.59 [a,b]
Total Cholesterol (mg/dL)	51.5 ± 8.80 [a]	60.0 ± 12.1 [a, b]	56.9 ± 14.1 [a]	71.7 ± 10.8 [b]
HDL Cholesterol (mg/dL)	18.0 ± 3.25 [a]	23.5 ± 3.94 [b]	19.9 ± 5.43 [a]	29.0 ± 2.69 [c]
Non-HDL Cholesterol (mg/dL)	33.5 ± 5.95 [a]	36.4 ± 9.09 [a]	37.0 ± 9.86 [a]	42.6 ± 8.87 [a]

Co: Control group; Co+Ly: Control group supplemented with lycopene; Ob: Obese group; Ob+Ly: Hypercaloric group supplemented with lycopene. HDL: high density lipoprotein. Data parametric expressed in mean ± standard deviation. Comparison by two-way ANOVA with Holm-Sidak post-hoc test. Different letters correspond to the significant statistical difference ($p < 0.05$); $n = 6$ animals/group.

3.4. Lycopene Ameliorates Obesity-Related Hepatic Steatosis

Figure 2 shows the determination of the hepatic steatosis by histological parameters. Independent of lycopene supplementation, the hepatic tissue of the obese animals (Ob and Ob+Ly) presented higher macrovesicular steatosis. Lycopene was efficient in attenuating the microvesicular steatosis of treated obese animals (Ob+Ly).

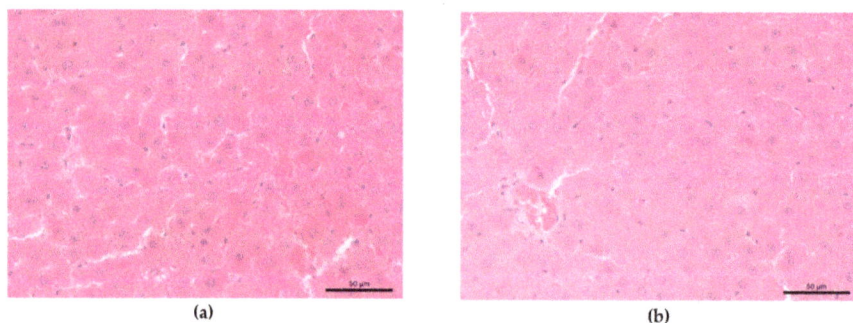

(a)

(b)

Figure 2. *Cont.*

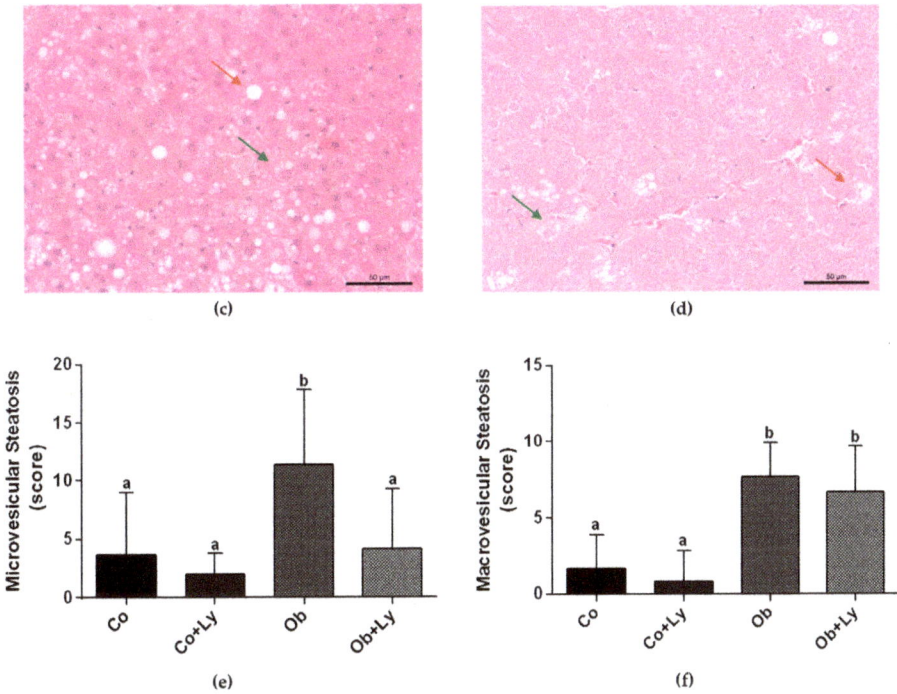

Figure 2. Determination of the hepatic steatosis by hepatic tissue histology stained with haematoxylin and eosin (H&E) (**a**) Illustrative picture (40× magnification) of the control group (Co); (**b**) illustrative picture (40× magnification) of the control group supplemented with lycopene group (Co+Ly); (**c**) illustrative picture (40× magnification) of the obese group (Ob); (**d**) illustrative picture (40× magnification) of the obese group supplemented with lycopene (Ob+Ly); (**e**) microvesicular steatosis (score); (**f**) macrovesicular steatosis (score). Presence of macro (red arrows) and microvesicular (green arrow) steatosis in obese groups. Data expressed in mean ± standard deviation. Comparison by Poasson distribution followed by the post-hoc test Wald multi-comparison. Different letters correspond to the significant statistical difference ($p < 0.05$); $n = 6$ animals/group.

3.5. Lycopene Ameliorates TNF-α Levels in Hepatic Tissue

In hepatic tissue, there was no difference between groups for the pro-inflammatory cytokine IL-6. The Ob group presented the highest levels of TNF-α and the supplementation with lycopene in obese animals (Ob+Ly) was effective in the attenuation of cytokine levels (Figure 3).

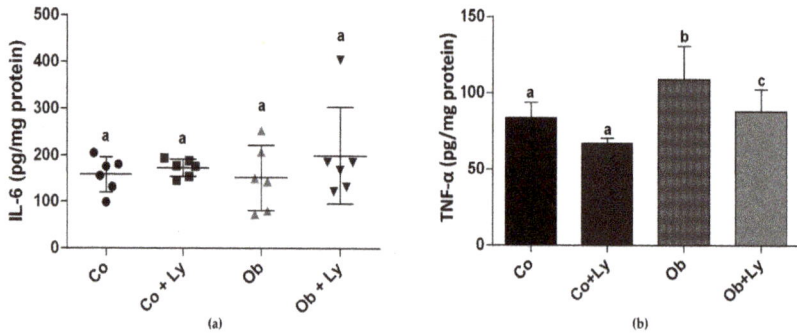

(a)   (b)

Figure 3. Inflammatory process in hepatic tissue: (**a**) IL-6 (pg/mg protein); (**b**) TNF-α (pg/mg protein). Co: Control group; Co+Ly: Control group supplemented with lycopene; Ob: Obese group; Ob+Ly: Hypercaloric group supplemented with lycopene. Data parametric expressed in mean ± standard deviation. Comparison by two-way ANOVA with Holm-Sidak post-hoc test. Data non-parametric expressed in median and interquartile range. Comparison by Kruskall Wallis test with Tukey *post-hoc* test. Different letters correspond to the significant statistical difference ($p < 0.05$); $n = 6$ animals/group. IL-6: Interleukin-6; TNF-α: Tumor necrosis factor-α.

3.6. Lycopene Shows Anti-Lipid Peroxidation Activity in Hepatic Tissue

The lipid peroxidation represented by MDA formation in the liver was higher in the control group. Lycopene was effective in diminished oxidative damage to lipids since it is possible to observe the lower levels of MDA in supplemented groups (Co+Ly and Ob+Ly) (Figure 4).

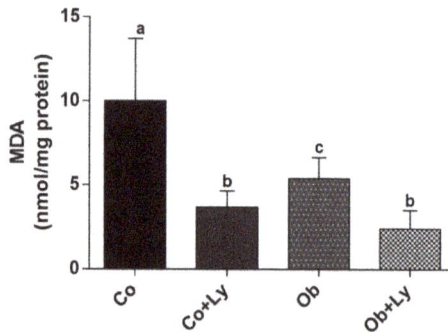

Figure 4. Assessment of lipid peroxidation through the quantification of malondialdehyde (MDA) in the liver: MDA (nmol/mg protein). Co: Control group; Co+Ly: Control group supplemented with lycopene; Ob: Obese group; Ob+Ly: Hypercaloric group supplemented with lycopene. Data parametric expressed in mean ± standard deviation. Comparison by two-way ANOVA with Holm-Sidak post-hoc. Different letters correspond to the significant statistical difference ($p < 0.05$); $n = 6$ animals/group.

3.7. Lycopene Improves Antioxidant Enzyme Activity in Hepatic Tissue

Figure 5 shows the antioxidant enzyme activity in the liver. Obese groups (Ob and Ob+Ly) showed reduced activity of CAT and the lycopene supplementation (Ob+Ly) improved CAT activity. The lycopene also had a positive effect on the SOD activity of the Ob+Ly group, which reached similar activity to the control groups (Co and Co+Ly).

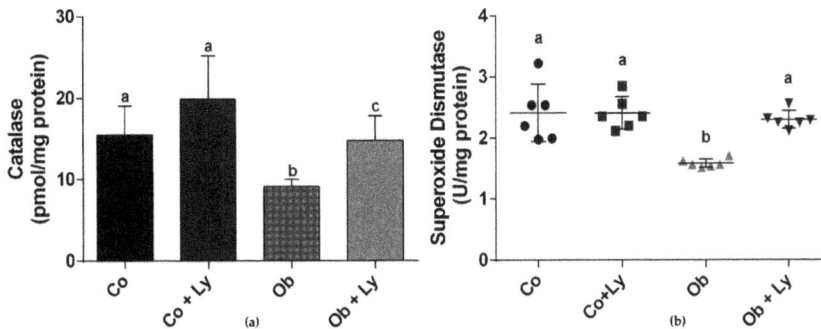

Figure 5. Antioxidant enzyme activity: (**a**) Catalase (pmol/mg protein); (**b**) Superoxide Dismutase (U/mg protein). Co: Control group; Co+Ly: Control group supplemented with lycopene; Ob: Obese group; Ob+Ly: Hypercaloric group supplemented with lycopene. Data parametric expressed in mean ± standard deviation. Comparison by two-way ANOVA with Holm-Sidak post-hoc test. Data non-parametric expressed in median and interquartile range. Comparison by Kruskall Wallis test with Tukey post-hoc test. Different letters correspond to the significant statistical difference ($p < 0.05$); $n = 6$ animals/group.

3.8. Lycopene Improves Total Antioxidant Capacity in Hepatic Tissue of Obese Rats

Figure 6 shows the total hepatic antioxidant capacity. The antioxidant protection was higher in the obese group supplemented with lycopene (Ob+Ly).

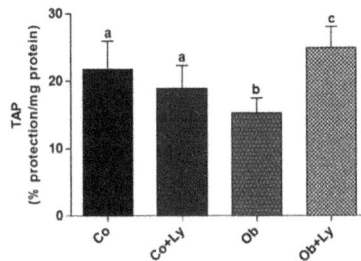

Figure 6. Total antioxidant performance test (TAP) (% protection/mg protein). Co: Control group; Co+Ly: Control group supplemented with lycopene; Ob: Obese group; Ob+Ly: Hypercaloric group supplemented with lycopene. Data parametric expressed in mean ± standard deviation. Comparison by two-way ANOVA with Holm-Sidak post-hoc. Different letters correspond to the significant statistical difference ($p < 0.05$); $n = 6$ animals/group.

4. Discussion

The aim of this study was to investigate the therapeutic potential of lycopene on the pathophysiological processes of NAFLD in obese animals fed a hypercaloric diet. The experimental model proposed in this study promoted metabolic and hepatic changes, represented by increased levels of fasting blood glucose and of plasmatic and hepatic triglycerides both associated with the accumulation of fat in the liver. This hepatic disturbance could be observed in the histological analysis through macro and microvesicular steatosis. However, there were no changes in liver enzymes ALT and AST evidencing that there was no progression of the disease. Although the model did not transpose steatosis, the obese animals (Ob) presented high levels of TNF-α levels, a potent pro-inflammatory cytokine. A greater weight gain and a different body composition represented by the adiposity index was also observed. The animals, which were fed a hypercaloric diet, ate a smaller amount of chow in grams. This event did not interfere with the caloric intake value as these animals received a higher caloric amount derived from water intake containing 25% sucrose. Despite the similar caloric intake between the groups, the difference in diet composition, in other words, its macronutrient quality and distribution reflected a greater weight gain in the groups which received the hypercaloric diet, in addition to all metabolic changes remarked.

Supplementation with lycopene did not influence the chow intake behavior and the weight gain of the treated groups. Although several studies have listed lycopene as an inducer of adiponectin concentrations [23,31], and adipokine, where serum levels are negatively associated with obesity by increasing energy expenditure [32], few have signed lycopene as a subsidiary in weight control [31]. The lycopene also did not influence the fasting blood glucose, although the action of lycopene in glycemic control of diabetic animals was found in the literature [33,34].

Lycopene is absorbed in the intestine and stored in addition to other organs in the liver where it exerts its biological activities [35]. Thus, the lycopene acts as a protective factor in the organ due to its antioxidant effect [36]. However, although redox imbalance state plays an important role in the pathogenesis of NAFLD, other metabolic disturbances are important as changes in energy metabolism [35]. Accordingly, the positive effects of lycopene were not only restricted to the protective effect in the antioxidant hepatic system, discussed later, one of the most relevant effects of lycopene supplementation was its action on lipid metabolism.

The liver is responsible for several functions in the body, such as cholesterol, fatty acids, carbohydrates, and protein metabolism, besides the formation of plasma proteins, and production of bile among other processes [37]. Some members of the nuclear receptor (NR) family are responsible for regulating these mechanisms in order to maintain the hepatic and body homeostasis. These receptors form heterodimers with the retinoid X receptor (RXR), resulting in gene transcription [38]. The heterodimers may be activated by specific ligands for each component of the pair or by simultaneous binding. Lycopene seems to act as an RXR ligand and is, therefore, capable of activating several NRs, although it has an important affinity for peroxisome proliferator-activated receptors (PPARs) family [39]. Anyway, these receptors are highly expressed in the liver and are involved in the transcription of enzymes related to lipid metabolism. Although all have important metabolic functions the PPAR-α is identified as the master regulator of hepatic lipid metabolism [40]. Its expression is inversely proportional to the progression of the histological severity of NAFLD and directly proportional to the improvement of the inflammatory state [41]. PPAR-α plays a key role in lipid catabolism through regulating target genes involved in β-oxidation, uptake and activation of fatty acids, and lipolysis [42]. Thus, PPAR-α has been targeted in the development of drugs for the NAFLD treatment [43], and lycopene was shown to have a protective effect on the disease through its binding in these NR [44].

Histologically, NAFLD is characterized by macrovesicular steatosis, that is, large or small lipid droplets are present in the cytoplasm with nucleus displacement [45]. However, approximately 10% of the liver biopsies of patients with NALFD present microvesicular steatosis [46], which is characterized by the accumulation of innumerable lipid droplets with a centrally placed nucleus [47]. The extensive microvesicular steatosis is associated with alcoholic fatty liver disease [48]. The difference between macrovesicular and microvesicular steatosis is not restricted to the histological aspect. Macrovesicular steatosis when alone is associated with a good prognosis with rare progression to fibrosis or cirrhosis [46]. Contrary, the microvesicular steatosis is a serious condition related to fibrosis, cholestasis, necrosis [49], and moreover to an impairment of the mitochondrial fatty acid oxidation [50]. The details surrounding the formation of these lipid droplets remain to be defined [51], and this study is not able to discern the relationship between micro and macrovesicular steatosis.

Interestingly, in the present study, lycopene was able to reduce microvesicular steatosis. This effect could be attributed to enhanced β-oxidation, an impaired pathway in this pattern of lipid accumulation [50]. Gala Martín-Pozuelo and colleagues [52] followed for seven weeks 24 Sprague-Dawley rats fed a hypercaloric diet and provided with water or tomato juice. The researchers concluded that the lycopene increased the activity of some enzymes involved in β-oxidation in the liver of animals that received the juice. This event was attributed to the activation of PPAR-α by lycopene.

Several researchers with animal models have investigated the therapeutic effect of lycopene on blood lipids [53–56]. In this study, an increase in total cholesterol was observed in the groups supplemented with lycopene. Such a finding may be justified by the higher levels of HDL in these animals since there was no change in non-HDL levels. HDL is the lipoprotein responsible for the reverse transport of cholesterol by removing cholesterol from the peripheral vessels and directing it to the liver. This added to other properties such as antioxidative, anti-inflammatory, vasodilatory, antithrombotic, and cytoprotective, which makes HDL a protective factor for cardiovascular diseases intrinsically associated with obesity [57]. In this sense, pharmacological and dietary therapies have been investigated with the objective of elevating HDL [58]. This lipoprotein beyond cholesterol is responsible for the transport of other molecules, including proteins, small RNAs, bioactive lipids, hormones, vitamins, and carotenoids. Hence, it has been suggested that the content of carotenoids plays a role in protecting HDL from oxidative modification [59].

Among the reasons for high HDL levels is the increase in adipose tissue observed in obesity, which is characterized by a chronic inflammation state of low grade. Serum amyloid A (SAA) is one of these pro-inflammatory proteins, which in high concentrations displaces apoprotein AI (apoA-I) as the predominantly apolipoprotein of HDL, making the lipoprotein dysfunctional [60]. McEneny and colleagues [61] studied the effects of lycopene on HDL and SAA in moderately overweight middle-aged men and women. Lycopene intervention diminished inflammation by decreasing systemic levels of SAA, hence reducing the association of SAA with HDL, and it even positively influenced the structural and/or functional composition of the apo AI. In this way, the carotenoid helped to restore the HDL functions. In addition, HDL is produced in the liver and intestine and its formation is controlled by NR [62]. PPAR-α is able to increase the synthesis of apoAI and the ATP-binding cassette A1 transporter (ABCA1) expression. In the intestine, activation of PPAR-α is facilitated by the binding of agonists present in the diet, such as lycopene. The action of lycopene in both the liver and the intestine results in increased plasma HDL levels [63]. Further, the agonist effect of lycopene on LXR also seems to increase the expression of ABCA1 and hence increase HDL levels [64].

The inflammation and oxidative stress are two relevant features of NAFLD. The liver inflammatory process represents an important pathway in the development of NAFLD/NASH, therefore, research has sought its control [65]. The hypercaloric diet increases the hepatic levels of the pro-inflammatory cytokine TNF-α, probably through activation of nuclear factor kappa B (NF-κB) signaling [66]. It is established in the literature that the anti-inflammatory potential of lycopene by promoting a significant inhibition of the NF-κB is probably mediated by PPARs [67]. In the present study, lycopene reduced the levels of TNF-α, reflecting control of the pro-inflammatory mechanisms, and also contributing to a reduction in pattern and extent of liver damage [68]. The inflammatory response may promote ROS production by the immune system cells, which contributes to liver damage [69]. Furthermore, ROS are able to activate NF-κB to induce pro-inflammatory cytokines contributing to this vicious cycle between inflammation and oxidative stress [14].

The accumulation of hepatic lipid related to a hypercaloric diet induces a metabolic shift in order to overcome this amount of lipid. This shift includes improvements in β-oxidation, Krebs cycle, and stimulation of oxidative phosphorylation (OXPHOS) [70]. However, with the chronic overnutrition, the mitochondrial adaptation is insufficient, leading to ROS overproduction [14]. One of the consequences of the redox imbalance is the lipid lipoperoxidation represented here by the formation of MDA. The MDA levels were higher in the control groups due to the lipid source of diet since the polyunsaturated fatty acids (PUFAs) present in soybean oil are more susceptible to lipid peroxidation [71]. Nevertheless, lycopene supplementation elevated SOD and CAT activity and significantly decreased the MDA level in hepatic tissue, probably due to nuclear factor erythroid 2–related factor 2 (Nrf2) activation. The Nrf2 system plays a crucial role in the regulation of redox homeostasis by orchestrating the antioxidant defense mechanisms by inducing endogenous antioxidant enzymes [72]. In addition, lycopene was efficient in increasing total antioxidant capacity in the obese group supplemented with lycopene (Ob+Ly), contributing to the control of pro-oxidative and pro-inflammatory processes.

Despite the outcome of lycopene effects on NAFLD observed in this study, there are some limitations regarding the translation of this study to humans. The lycopene concentration used in the treatment is higher than that could be offered by a balanced diet. More studies are needed to determine the lycopene effects on NAFLD in humans as well as the appropriated dose for treatment.

5. Conclusions

In summary, lycopene decreases plasmatic and hepatic triglycerides levels, increased HDL-c, mitigates microvesicular steatosis, attenuates levels of the pro-inflammatory cytokine, and contributes to the redox balance in the liver (Figure 7) In view of these results, it may be concluded that lycopene modulates important pathophysiological processes of the NAFLD, being a therapeutic potential for disease control.

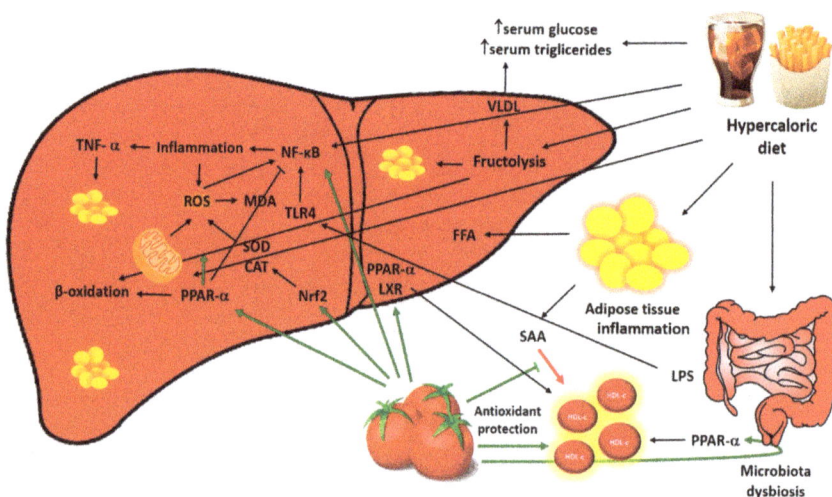

Figure 7. Role of lycopene in the pathophysiological process of non-alcoholic fatty liver disease (NAFLD). The hypercaloric diet provides large amounts of fat and especially sugars, which saturate the liver's ability to export triglycerides leading to its accumulation in the hepatocyte in addition to increased levels of triglyceride and plasma glucose. These macronutrients also activate inflammatory pathways directly by NF-κB activation and indirectly through alteration of intestinal permeability and translocation of LPS to the liver. The excess nutrients overload the oxidation capacity of mitochondria leading to ROS production, establishing a mutually dependent process between inflammation and oxidative stress. Lycopene was efficient in controlling the inflammatory and oxidative pathways, as well as in the antioxidant activity and assisted in the control of the accumulation of hepatic lipids, activating β-oxidation. Furthermore, the carotenoid improved blood lipids profile by lowering triglycerides and increasing HDL-C levels. LPS: Lipopolysaccharides; SAA: Serum amyloid A; HDL-c: High-density lipoprotein cholesterol; VLDL: Very low-density lipoprotein, FFA: Free fatty acid; Nrf2: Nuclear factor erythroid 2–related factor 2; PPAR-α: Peroxisome proliferator-activated receptor α; SOD: Superoxide dismutase CAT: Catalase; TLR4: Toll-like receptor 4; ROS: Reactive oxygen species; MDA: Malondialdehyde; TNF-α: Tumor necrosis factor-α.

Author Contributions: Conceptualization, F.M., C.R.C. and A.L.d.A.F.; methodology, J.L.G., A.J.T.F., F.V.F.-F., F.K.H., D.H.S.d.C., and M.R.C.; formal analysis, M.R.C., C.R.d.A.; writing—original draft preparation, M.R.C., C.S.G., and C.C.V.d.A.S.; writing—review and editing, M.R.C., J.L.G., C.R.C., F.M.; supervision, M.R.C. and F.M.; funding acquisition, A.L.d.A.F.

Funding: This research was supported by Conselho Nacional de Desenvolvimento Científico e Tecnológico—CNPq, grant number 42.42.09/2016-0).

Acknowledgments: Fundação de Amparo à Pesquisa do Estado de São Paulo (FAPESP).

Conflicts of Interest: The authors declare no conflict of interest.

References

1. World Health Organization. *Obesity: Preventing and Managing the Global Epidemic*; WHO Technical Report Series 894; World Health Organization: Geneva, Switzerland, 2000.
2. Kelly, T.; Yang, W.; Chen, C.S.; Reynolds, K.; He, J. Global burden of obesity in 2005 and projections to 2030. *Int. J. Obes.* **2008**, *32*, 1431–1437. [CrossRef] [PubMed]
3. World Health Organization (WHO). Obesity and Overweight. Available online: https://www.who.int/news-room/fact-sheets/detail/obesity-and-overweight (accessed on 7 December 2018).
4. Popkin, B.M. Nutrition transition and the global diabetes epidemic. *Curr. Diab. Rep.* **2015**, *15*, 1617–1622. [CrossRef] [PubMed]
5. Mehrabani, J.; Ganjifar, Z.K. Overweight and obesity: A brief challenge on prevalence, complications and physical activity among men and women. *Women's Health* **2018**, *7*, 19–24. [CrossRef]
6. DiNicolantonio, J.J.; Berger, A. Added sugars drive nutrient and energy deficit in obesity: A new paradigm. *Open Heart* **2016**, *3*, 1–6. [CrossRef] [PubMed]
7. Fleur, S.E.; Luijendijk, M.C.M.; van Rozen, A.J.; Kalsbeek, A.; Adan, R.A.H. A free-choice high-fat high-sugar diet induces glucose intolerance and insulin unresponsiveness to a glucose load not explained by obesity. *Int. J. Obes.* **2010**, *35*, 595–604. [CrossRef] [PubMed]
8. Segula, D. Complications of obesity in adults: A short review of the literature. *Malawi Med. J.* **2014**, *26*, 20–24. [PubMed]
9. Lu, F.B.; Hu, E.D.; Xu, L.M.; Chen, L.; Wu, J.L.; Li, H.; Chen, D.Z.; Chen, Y.P. The Relationship between obesity and the severity of non-alcoholic fatty liver disease: Systematic review and meta-analysis. *Expert Rev. Gastroenterol. Hepatol.* **2018**, *12*, 491–502. [CrossRef] [PubMed]
10. Lim, J.S.; Mietus-Snyder, M.; Valente, A.; Schwarz, J.M.; Lustig, R.H. The Role of fructose in the pathogenesis of nafld and the metabolic syndrome. *Nat. Rev. Gastroenterol. Hepatol.* **2010**, *7*, 251–264. [CrossRef]
11. Ni, Y.; Zhuge, F.; Nagashimada, M.; Ota, T. Novel action of carotenoids on non-alcoholic fatty liver disease: Macrophage polarization and liver homeostasis. *Nutrients* **2016**, *8*, 391. [CrossRef]
12. Shi, L.; Tu, B.P. Acetyl-CoA and the regulation of metabolism: Mechanisms and consequences. *Curr. Opin. Cell Biol.* **2015**, *33*, 125–131. [CrossRef]
13. Kim, C.W.; Addy, C.; Kusunoki, J.; Anderson, N.N.; Deja, S.; Fu, X.; Burgess, S.C.; Li, C.; Chakravarthy, M.; Previs, S.; et al. Acetyl-CoA carboxylase inhibition reduces hepatic steatosis but elevates plasma triglycerides in mice and humans: A bedside to bench investigation. *Cell Metab.* **2017**, *26*, 394–406. [CrossRef] [PubMed]
14. Simões, I.C.M.; Fontes, A.; Pinton, P.; Zischka, H.; Wieckowski, M.R. Mitochondria in non-alcoholic fatty liver disease. *Int. J. Biochem. Cell Biol.* **2018**, *95*, 93–99. [CrossRef] [PubMed]
15. Nita, M.; Grzybowski, A. The Role of the reactive oxygen species and oxidative stress in the pathomechanism of the age-related ocular diseases and other pathologies of the anterior and posterior eye segments in adults. *Oxid. Med. Cell. Longev.* **2016**, *2016*, 3164734. [CrossRef] [PubMed]
16. Fritsche, K.L. The science of fatty acids and inflammation. *Adv. Nutr.* **2015**, *6*, 293–301. [CrossRef] [PubMed]
17. Lau, E.; Carvalho, D.; Freitas, P. Gut microbiota: Association with NAFLD and metabolic disturbances. *BioMed Res. Int.* **2015**, *2015*, 979515. [CrossRef] [PubMed]
18. Sumida, Y.; Yoneda, M. Current and future pharmacological therapies for NAFLD/NASH. *J. Gastroenterol.* **2018**, *53*, 362–376. [CrossRef] [PubMed]
19. Ferramosca, A.; Di Giacomo, M.; Zara, V. Antioxidant dietary approach in treatment of fatty liver: New insights and updates. *World J. Gastroenterol.* **2017**, *23*, 4146–4157. [CrossRef]

20. Kawata, A.; Murakami, Y.; Suzuki, S.; Fujisawa, S. Anti-inflammatory activity of β-carotene, lycopene and tri-n-butylborane, a scavenger of reactive oxygen species. *In Vivo* **2018**, *32*, 255–264.

21. Canadian Council on Animal Care. *Guide to the Care and Use of Experimental Animals*; CCAC: Ottawa, ON, Canada, 1993; Volume 1, p. 209.

22. Francisqueti, F.V.; Minatel, I.O.; Ferron, A.J.T.; Bazan, S.G.Z.; Silva, V.S.S.; Garcia, J.L.; Campos, D.H.S.; Ferreira, A.L.; Moreto, F.; Cicogna, A.C.; et al. Effect of gamma-oryzanol as therapeutic agent to prevent cardiorenal metabolic syndrome in animals submitted to high sugar-fat diet. *Nutrients* **2017**, *9*, 1299. [CrossRef]

23. Luvizotto, R.A.M.; Nascimento, A.F.; Miranda, N.C.M.; Wang, X.D.; Ferreira, A.L.A. Lycopene-rich tomato oleoresin modulates plasma adiponectin concentration and mRNA levels of adiponectin, SIRT1, and FoxO1 in adipose tissue of obese Rats. *Hum. Exp. Toxicol.* **2015**, *34*, 612–619. [CrossRef]

24. Luvizotto, R.D.A.M.; Nascimento, A.F.; Imaizumi, E.; Pierine, D.T.; Conde, S.J.; Correa, C.R.; Yeum, K.J.; Ferreira, A.L.A. Lycopene supplementation modulates plasma concentrations and epididymal adipose tissue mRNA of leptin, resistin and IL-6 in diet-induced obese rats. *Br. J. Nutr.* **2013**, *110*, 1803–1809. [CrossRef] [PubMed]

25. Ferreira, A.L.; Salvadori, D.M.; Nascimento, M.C.; Rocha, N.S.; Correa, C.R.; Pereira, E.J.; Matsubara, L.S.; Matsubara, B.B.; Ladeira, M.S. Tomato-oleoresin supplement prevents doxorubicin-induced cardiac myocyte oxidative DNA damage in rats. *Mutat. Res.* **2007**, *631*, 26–35. [CrossRef] [PubMed]

26. Liang, W.; Menke, A.L.; Driessen, A.; Koek, G.H.; Lindeman, J.H.; Stoop, R.; Havekes, L.M.; Kleemann, R.; Van Den Hoek, A.M. Establishment of a general NAFLD scoring system for rodent models and comparison to human liver pathology. *PLoS ONE* **2014**, *9*, e115922. [CrossRef] [PubMed]

27. Moreto, F.; de Oliveira, E.P.; Manda, R.M.; Burini, R.C. The higher plasma malondialdehyde concentrations are determined by metabolic syndrome-related glucolipotoxicity. *Oxid. Med. Cell. Longev.* **2014**, *2014*, 505368. [CrossRef] [PubMed]

28. Marklund, S.L. Product of extracellular-superoxide dismutase catalysis. *FEBS Lett.* **1985**, *184*, 237–239. [CrossRef]

29. Aebi, H. Oxygen Radicals in Biological Systems: Preface. *Methods Enzymol.* **1990**, *186*, 121–126.

30. Beretta, G.; Aldini, G.; Facino, R.M.; Russell, R.M.; Krinsky, N.I.; Yeum, K.J. Total antioxidant performance: A validated fluorescence assay for the measurement of plasma oxidizability. *Anal. Biochem.* **2006**, *354*, 290–298. [CrossRef] [PubMed]

31. Li, Y.F.; Chang, Y.Y.; Huang, H.C.; Wu, Y.C.; Yang, M.D.; Chao, P.M. Tomato juice supplementation in young women reduces inflammatory adipokine levels independently of body fat reduction. *Nutrition* **2014**, *31*, 691–696. [CrossRef]

32. Liu, M.; Liu, F. Transcriptional and post-translational regulation of adiponectin. *Biochem. J.* **2009**, *425*, 41–52. [CrossRef]

33. Soleymaninejad, M.; Joursaraei, S.G.; Feizi, F.; Jafari Anarkooli, I. The effects of lycopene and insulin on histological changes and the expression level of Bcl-2 family genes in the hippocampus of streptozotocin-induced diabetic rats. *J. Diabetes Res.* **2017**, *2017*, 4650939. [CrossRef]

34. Aydin, M.; Çelik, S. Effects of lycopene on plasma glucose, insulin levels, oxidative stress, and body weights of streptozotocin-induced diabetic rats. *Turk. J. Med. Sci.* **2012**, *42*, 1406–1413.

35. Bernal, C.; Martín-Pozuelo, G.; Lozano, A.B.; Sevilla, Á.; García-Alonso, J.; Canovas, M.; Periago, M.J. Lipid biomarkers and metabolic effects of lycopene from tomato juice on liver of rats with induced hepatic steatosis. *J. Nutr. Biochem.* **2013**, *24*, 1870–1881. [CrossRef]

36. Yilmaz, B.; Sahin, K.; Bilen, H.; Bahcecioglu, I.H.; Bilir, B.; Ashraf, S.; Halazun, K.J.; Kucuk, O. Carotenoids and non-alcoholic fatty liver disease. *Hepatobiliary Surg. Nutr.* **2015**, *4*, 161–171.

37. Corless, J.K.; Middleton, H.M. Normal liver function. A basis for understanding hepatic disease. *Arch. Intern. Med.* **1983**, *143*, 2291–2294. [CrossRef]

38. Evans, R.M.; Mangelsdorf, D.J. Nuclear receptors, RXR the big bang. *Cell* **2014**, *157*, 255–266. [CrossRef]

39. Caris-Veyrat, C.; Garcia, A.L.; Reynaud, E.; Lucas, R.; Aydemir, G.; Rühl, R. A review about lycopene-induced nuclear hormone receptor signalling in inflammation and lipid metabolism via still unknown endogenous Apo-10′-lycopenoids. *Int. J. Vitam. Nutr. Res.* **2016**, *86*, 62–70. [CrossRef]

40. Desvergne, B.; Wahli, W. Peroxisome proliferator-activated receptors: Nuclear control of metabolism. *Endocr. Rev.* **2004**, *20*, 649–688.

41. Prawitt, J.; Caron, S.; Van Hul, W.; Lefebvre, P.; Verrijken, A.; Van Gaal, L.; Michielsen, P.; Hubens, G.; Mertens, I.; Taskinen, M.R.; et al. PPARα gene expression correlates with severity and histological treatment response in patients with non-alcoholic steatohepatitis. *J. Hepatol.* **2015**, *63*, 164–173.

42. Rakhshandehroo, M.; Knoch, B.; Müller, M.; Kersten, S. Peroxisome proliferator-activated receptor alpha target genes. *PPAR Res.* **2010**, *2010*, 612089. [CrossRef]

43. Pappachan, J.M.; Babu, S.; Krishnan, B.; Ravindran, N.C. Non-alcoholic fatty liver disease: A clinical update. *J. Clin. Transl. Hepatol.* **2017**, *5*, 384–393. [CrossRef]

44. Jiang, W.; Guo, M.H.; Hai, X. Hepatoprotective and antioxidant effects of lycopene on non-alcoholic fatty liver disease in rat. *World J. Gastroenterol.* **2016**, *22*, 10180–10188. [CrossRef]

45. Brunt, E.M.; Tiniakos, D.G. Histopathology of nonalcoholic fatty liver disease. *World J. Gastroenterol.* **2010**, *16*, 5286–5296. [CrossRef]

46. Tandra, S.; Yeh, M.M.; Brunt, E.M.; Vuppalanchi, R.; Cummings, O.W.; Ünalp-Arida, A.; Wilson, L.A.; Chalasani, N. Presence and significance of microvesicular steatosis in nonalcoholic fatty liver disease. *J. Hepatol.* **2010**, *55*, 654–659. [CrossRef]

47. Benedict, M.; Zhang, X. Non-alcoholic fatty liver disease: An expanded review. *World J. Hepatol.* **2017**, *9*, 715–732. [CrossRef]

48. Seitz, H.K.; Bataller, R.; Cortez-Pinto, H.; Gao, B.; Gual, A.; Lackner, C.; Mathurin, P.; Mueller, S.; Szabo, G.; Tsukamoto, H. Alcoholic Liver Disease. *Nat. Rev.* **2018**, *4*, 109–120. [CrossRef]

49. Fromenty, B.; Pessayre, D. Impaired mitochondrial function in microvesicular steatosis. *J. Hepatol.* **1997**, *26*, 43–53. [CrossRef]

50. Fromenty, B.; Berson, A.; Pessayre, D. Microvesicular steatosis and steatohepatitis: Role of mitochondrial dysfunction and lipid peroxidation. *J. Hepatol.* **2005**, *26*, 13–22. [CrossRef]

51. Mashek, D.G.; Khan, S.A.; Sathyanarayan, A.; Ploeger, J.M.; Franklin, M.P. Hepatic Lipid droplet biology: Getting to the root of fatty liver. *Hepatology* **2015**, *62*, 964–967. [CrossRef]

52. Martín-Pozuelo, G.; Santaella, M.; Hidalgo, N.; García-Alonso, J.; Periago, M.J.; Navarro-González, I.; Ros, G.; González-Barrio, R.; Gómez-Gallego, C. The effect of tomato juice supplementation on biomarkers and gene expression related to lipid metabolism in rats with induced hepatic steatosis. *Eur. J. Nutr.* **2014**, *54*, 933–944. [CrossRef]

53. Hu, M.Y.; Li, Y.-L.; Jiang, C.H.; Liu, Z.Q.; Qu, S.L.; Huang, Y.M. Comparison of lycopene and fluvastatin effects on atherosclerosis induced by a high-fat diet in rabbits. *Nutrition* **2008**, *24*, 1030–1038. [CrossRef]

54. Lorenz, M.; Fechner, M.; Kalkowski, J.; Fröhlich, K.; Trautmann, A.; Böhm, V.; Liebisch, G.; Lehneis, S.; Schmitz, G.; Ludwig, A.; et al. Effects of lycopene on the initial state of atherosclerosis in New Zealand white (NZW) rabbits. *PLoS ONE* **2012**, *7*, e30808. [CrossRef]

55. Mulkalwar, S.A.; Munjal, N.S.; More, U.K.; More, B.; Chaudhari, A.B.; Dewda, P.R. Effect of purified lycopene on lipid profile, antioxidant enzyme and blood glucose in hyperlipidemic rabbits. *Am. J. Pharm. Tech. Res.* **2012**, *2*, 460–470.

56. Hassan, H.A.; Edrees, G.M. Therapeutic effect of lycopene-rich tomato juice on cardiac disorder in rats fed on fried food in oxidized frying oil. *Egypt. J. Hosp. Med.* **2004**, *14*, 115–126.

57. Kosmas, C.E.; Martinez, I.; Sourlas, A.; Bouza, K.V.; Campos, F.N.; Torres, V.; Montan, P.D.; Guzman, E. High-density lipoprotein (HDL) functionality and its relevance to atherosclerotic cardiovascular disease. *Drugs Context.* **2018**, *7*, 212525. [CrossRef]

58. Farrer, S. Beyond statins: Emerging evidence for hdl-increasing therapies and diet in treating cardiovascular disease. *Adv. Prev. Med.* **2018**, *2018*, 6024747. [CrossRef]

59. Dias, I.H.K.; Polidorib, M.C.; Li, L.; Weber, D.; Stahlb, W.; Nellese, G.; Gruned, T.; Griffiths, H.R. Plasma levels of HDL and carotenoids are lower in dementia patients with vascular comorbidities. *J. Alzheimer's Dis.* **2018**, *40*, 399–408. [CrossRef]

60. Zhao, Y.; He, X.; Shi, X.; Huang, C.; Liu, J.; Zhou, S.; Heng, C.K. Association between serum amyloid a and obesity: A meta-analysis and systematic review. *Inflamm. Res.* **2010**, *59*, 323–334. [CrossRef]

61. McEneny, J.; Wade, L.; Young, I.S.; Masson, L.; Duthie, G.; McGinty, A.; McMaster, C.; Thies, F. Lycopene intervention reduces inflammation and improves hdl functionality in moderately overweight middle-aged individuals. *J. Nutr. Biochem.* **2013**, *24*, 163–168. [CrossRef]

62. Brunham, L.R.; Kruit, J.K.; Iqbal, J.; Fievet, C.; Timmins, J.M.; Pape, T.D.; Coburn, B.A.; Bissada, N.; Staels, B.; Groen, A.K.; et al. Intestinal ABCA1 directly contributes to HDL biogenesis in vivo. *J. Clin. Investig.* **2006**, *116*, 1052–1062. [CrossRef]

63. Colin, S.; Briand, O.; Touche, V.; Wouters, K.; Baron, M.; Pattou, F.; Hanf, R.; Tailleux, A.; Chinetti, G.; Staels, B.; et al. Activation of intestinal peroxisome proliferator-activated receptor- increases high-density lipoprotein production. *Eur. Heart J.* **2012**, *34*, 2566–2574. [CrossRef]

64. Hussain, M.M. Intestinal lipid absorption and lipoprotein formation. *Curr. Opin. Lipidol.* **2014**, *25*, 200–206. [CrossRef]

65. Sun, X.; Zhang, Y.; Xie, M. The role of peroxisome proliferator-activated receptor in the treatment of non-alcoholic fatty liver disease. *Acta Pharm.* **2017**, *67*, 1–13. [CrossRef]

66. Carlsen, H.; Haugen, F.; Zadelaar, S.; Kleemann, R.; Kooistra, T.; Drevon, C.A.; Blomhoff, R. Diet-induced obesity increases NF-KB signaling in reporter mice. *Genes Nutr.* **2009**, *4*, 215–222. [CrossRef]

67. Scirpo, R.; Fiorotto, R.; Villani, A.; Amenduni, M.; Spirili, C.; Strazzabosco, M. Stimulation of nuclear receptor ppar-γ limits nf-kb-dependent inflammation in mouse cystic fibrosis biliary epithelium. *Hepatology* **2016**, *62*, 1551–1562. [CrossRef]

68. Seo, Y.Y.; Cho, Y.K.; Bae, J.C.; Seo, M.H.; Park, S.E.; Rhee, E.J.; Park, C.Y.; Oh, K.W.; Park, S.W.; Lee, W.Y. Tumor necrosis factor-α as a predictor for the development of nonalcoholic fatty liver disease: A 4-year follow-up study. *Endocrinol. Metab.* **2013**, *28*, 41–45. [CrossRef]

69. Jaeschke, H. Reactive oxygen and mechanisms of inflammatory liver injury: Present concepts. *J. Gastroenterol. Hepatol.* **2011**, *26*, 173–179. [CrossRef]

70. Sunny, N.E.; Parks, E.J.; Browning, J.D.; Burgess, S.C. Excessive hepatic mitochondrial TCA cycle and gluconeogenesis in humans with nonalcoholic fatty liver disease. *Cell Metab.* **2011**, *14*, 804–810. [CrossRef]

71. Haggag, M.S.; Elsanhoty, R.M.; Ramadan, M.F. Impact of dietary oils and fats on lipid peroxidation in liver and blood of albino rats. *Asian Pac. J. Trop. Biomed.* **2014**, *4*, 52–58. [CrossRef]

72. Sahin, K.; Orhan, C.; Tuzcu, M.; Sahin, N.; Hayirli, A.; Bilgili, S.; Kucuk, O. Lycopene activates antioxidant enzymes and nuclear transcription factor systems in heat-stressed broilers. *Poult. Sci.* **2016**, *95*, 1088–1095. [CrossRef]

antioxidants

MDPI

Article

Preparation of Retinoyl-Flavonolignan Hybrids and Their Antioxidant Properties

Christopher S. Chambers [1], David Biedermann [1], Kateřina Valentová [1], Lucie Petrásková [1], Jitka Viktorová [2], Marek Kuzma [1] and Vladimír Křen [1,*]

[1]   Laboratory of Biotransformation, Institute of Microbiology of the Czech Academy of Sciences, Vídeňská 1083, 14220 Prague, Czech Republic
[2]   Department of Biochemistry and Microbiology, University of Chemistry and Technology, Technická 5, 16628 Prague, Czech Republic
*   Correspondence: kren@biomed.cas.cz; Tel.: +420-296-442-510

Received: 27 June 2019; Accepted: 19 July 2019; Published: 23 July 2019

Abstract: Antioxidants protect the structural and functional components in organisms against oxidative stress. Most antioxidants are of plant origin as the plants are permanently exposed to oxidative stress (UV radiation, photosynthetic reactions). Both carotenoids and flavonoids are prominent antioxidant and anti-radical agents often occurring together in the plant tissues and acting in lipophilic and hydrophilic milieu, respectively. They are complementary in their anti-radical activity. This study describes the synthesis of a series of hybrid ester conjugates of retinoic acid with various flavonolignans, such as silybin, 2,3-dehydrosilybin and isosilybin. Antioxidant/anti-radical activities and bio-physical properties of novel covalent carotenoid-flavonoid hybrids, as well as various mixtures of the respective parent components, were investigated. Retinoyl conjugates with silybin—which is the most important flavonolignan in silymarin complex—(and its pure diastereomers) displayed better 1,1-diphenyl-2-picrylhydrazyl (DPPH) radical scavenging activity than both the parent compounds and their equimolar mixtures.

Keywords: carotenoids; retinol; retinoic acid; vitamin A; flavonolignans; silymarin; antioxidant; anti-radical; esterification; conjugate

1. Introduction

Antioxidants are substances that protect the structural and functional components in living organisms—especially proteins, lipid structures (membranes) and nucleic acids—against oxidative stress. Protein damage by oxidative radicals and reactive oxygen species (ROS, e.g., $OH^\bullet$, $O_3^\bullet$, NO, $ROO^\bullet$, H_2O_2, $^1O_2$, HClO, etc.) leads to partial or complete loss of function and also to the risk of autoimmune diseases due to cross-immune reactivity between the damaged (immunogenic) and intact proteins [1]. Most of the antioxidants are of plant origin as the plants are permanently exposed to oxidative stress (UV radiation, photosynthetic reactions) and cannot escape [2]. Antioxidants mostly act in complex systems in biological structures, often as redox tandems. Neutralizing a highly reactive radical with an antioxidant can lead to another, usually less reactive radical, which can also be dangerous in the case of accumulation. Antioxidants—especially at higher doses—often act as oxidative stressors. Different antioxidant mechanisms of complementary antioxidants thus act much more effectively than single antioxidants even at a higher concentration. Cocktails of various antioxidants are often the key ingredients of various nutraceuticals claiming to protect against oxidative stress. The most important antioxidants include a number of vitamins (ascorbate, retinol and tocopherol) and prominent bioactives such as carotenoids and flavonoids [3]. Nevertheless, the concept of antioxidants as active redox species in vivo has been re-evaluated as a number of "antioxidants" actually act as weak prooxidants or they directly interact with some nuclear receptors, such as NF-κB, thus regulating the production of intracellular enzymatic antioxidant systems [4].

Carotenoids [5] and flavonoids [6] are natural products that are found throughout the plant kingdom and play a crucial role in photoreception, photoprotection, antioxidant processes and some of them—namely carotenoids—in photosynthesis [2]. They are also very important as exogenous biologically active substances (vitamins) for animals and humans and are commonly consumed in the diet (Figure 1).

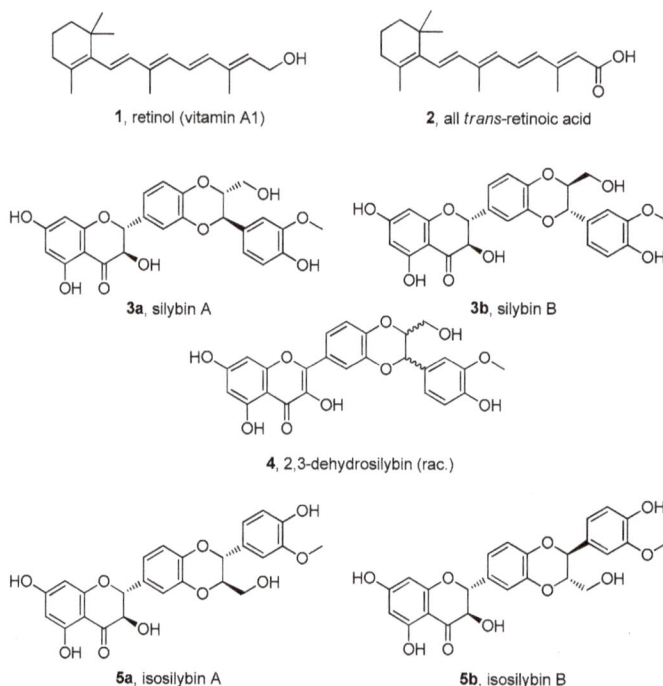

Figure 1. Structures of retinol, retinoic acid and selected flavonolignans from *Silybum marianum* (L.) Gaertn. (Asteraceae). Note to the numbering: Natural silybin is extracted and often used as an approximately equimolar diastereomeric mixture of silybin A (**3a**) and silybin B (**3b**). Diastereomeric mixtures of silybin and its derivatives will be denoted with the respective number and letters **ab** (for example natural mixture of **3a** and **3b** will be denoted **3ab**).

Combining antioxidants having different mechanisms of action in a single molecule often leads to a strong potentiation of the antioxidant activity. This approach has already been used in hybrids containing, for example, tocopherol and procaine [7]; iron chelator deferiprone (3-hydroxy-2-methyl-4(1*H*)-pyridinone) and 3,5-disubstituted-4-hydroxyphenyl derivatives like BHT (butylated hydroxytoluene) [8], flavonolignans and fatty acids [9] or ascorbic acid [10] with improved antioxidant or other specific properties. Carotenoids have been rarely used in this manner probably due to their limited stability. Astaxanthin has been esterified with ferulic acid to obtain astaxanthin diferulate, improving the radical scavenging ability [11]. 4′-Hydroxyflavone has also been combined with 12′-apo-β-carotenal, which resulted in the formation of C–C linked artificial carotenylflavonoid [12,13]. Flavonoids and especially their glycosides, are considerably more hydrophilic than carotenoids. Interactions of flavonoids with carotenoids often occur at the water/lipid interfaces. The activity of flavonoids appears to be more pronounced in the aqueous phase, whereas carotenoids are more active in the lipid phase [14].

The aim of the present study was to synthetize a series of hybrid conjugates of retinol (**1**) and/or retinoic acid (**2**) with various flavonolignans such as silybin A (**3a**), silybin B (**3b**), 2,3-dehydrosilybin (**4**), isosilybin A (**5a**) and isosilybin B (**5b**) from silymarin. (Figure 1) The biophysical properties of supramolecular conjugates of these two important types of antioxidants were investigated.

2. Materials and Methods

2.1. Chemicals

Retinol (**1**) and retinoic acid (**2**) were obtained from Sigma-Aldrich (Prague, Czech Republic). Silybin diastereomeric mixture (**3ab**), optically pure silybins A, B (**3a**, **3b**) [9] and isosilybin A (**5a**) [15] were prepared from silymarin purchased from Liaoning Senrong Pharmaceutical Co. (Panjin, People's Republic of China, batch no. 120501) according to the published procedures. 2,3-Dehydrosilybin (**4**) was prepared from silybin as previously described [16].

2.2. Nuclear Magnetic Resonance (NMR) and Mass Spectrometry (MS) Methodology

A Bruker Avance III 700 MHz spectrometer (700.13 MHz for ^{1}H, 176.05 MHz for ^{13}C) and Bruker Avance III 600 MHz spectrometer (600.23 MHz for ^{1}H, 150.93 MHz for ^{13}C, Bruker BioSpin, Rheinstetten, Germany) were used for NMR analysis of samples in dimethylsulfoxide (DMSO)-d_6 (99.8% atom D, VWR International, Stříbrná Skalice, Czech Republic) at 30 °C. The residual signals of the solvent were used as internal standards (δ_H 2.499, δ_C 39.46). The following experiments were performed and processed using the manufacturer's software (Topspin 3.2, Bruker BioSpin, Rheinstetten, Germany): ^{1}H NMR, ^{13}C NMR, ge-2D homonuclear correlation spectroscopy (COSY), ge-2D multiplicity-edited ^{1}H–^{13}C heteronuclear single-quantum correlation spectroscopy (HSQC), ge-2D ^{1}H–^{13}C heteronuclear multiple-bond correlation spectroscopy (HMBC) and 1D total correlation spectroscopy (TOCSY). ^{1}H NMR and ^{13}C NMR spectra were zero filled to four-fold data points and multiplied by a window function before the Fourier transformation.

The signal-to-noise ratio in ^{13}C NMR spectra was improved by application of a line broadening (1 Hz). The resolution in ^{1}H NMR spectra were accomplished by the two-parameter double-exponential Lorentz–Gauss function applied prior to Fourier transformation. Multiplicity-edited ^{1}H-^{13}C HSQC spectra were utilized to identify the multiplicity of the ^{13}C signals. Chemical shifts are reported in δ-scale. The digital resolution of spectra allowed us to report carbon chemical shift to two decimal places and hydrogen chemical shifts to three decimal places.

The LTQ Orbitrap XL hybrid mass spectrometer (Thermo Fisher Scientific, Waltham, MA, USA) equipped with an electrospray ion source was used to measure high resolution mass spectra (HRMS). The samples dissolved in MeOH were introduced into the mobile phase flow (MeOH/H$_2$O 4:1; 100 μL/min) using a 2 μL loop. Spray, capillary and tube lens voltage were 4.0 kV, −16 V, −120 V, respectively; capillary temperature was 275 °C.

2.3. HPLC

The Shimadzu Prominence System (Shimadzu Corporation, Kyoto, Japan) consisting of a DGU-20A3 mobile phase degasser, two LC-20AD solvent delivery units, a SIL-20AC cooling auto sampler, a CTO-10AS column oven and SPD-M20A diode array detector was employed for analytical high performance liquid chromatography (HPLC) analyses. The Chromolith Performance RP-18e monolithic column (100 mm × 3 mm i.d., Merck, Darmstadt, Germany) coupled with a guard column (5 mm × 4.6 mm i.d., Merck, Darmstadt, Germany) was used. Mobile phase acetonitrile (phase A) and water (phase B) were used for the analyses; gradient: 0 80% B; 20 min 0% B, 22–24.5 min 80% B at 25 °C; the flow rate was 1.0 mL/min. The Photometric Diode Array (PDA) data were acquired in the 200–450 nm range and signals at 360 nm were extracted. Chromatographic data were collected and processed using Shimadzu Solution software (version 5.75 SP2, Shimadzu Corporation, Tokyo, Japan) at a rate of 40 Hz and detector time constant of 0.025 s.

Preparative HPLC separations were carried out with a Shimadzu system consisting of a LC-8A high-pressure pump with a SPD-20A dual wavelength detector (with preparative cell) and FRC-10A fraction collector. The system was connected to a PC using CBM-20A command module and controlled by a LabSolution 1.24 SP1 software suite (all Shimadzu, Kyoto, Japan). All preparative HPLC separations were performed using an ASAHIPAK GS-310 20F column (Shodex, Munich, Germany) at 25 °C with MeOH as mobile phase, flow rate 5 mL/min and detection at 254 and 369 nm.

2.4. Chemical Synthesis

2.4.1. General Procedures for the Synthesis of Conjugates

Method A: Retinoic acid (**2**, 0.5 mmol, 1 eq), flavonolignan (**3–5**, 0.75 mmol, 1.5 eq), N,N'-dicyclohexylcarbodiimide (DCC, 1 mmol, 2 eq) and 4-dimethylaminopyridine (DMAP, catalytic amount) were dissolved in dry tetrahydrofuran (THF, 25 mL). The reaction mixture was stirred at room temperature (25 °C) in darkness under argon for 18 h. The solvent was removed in vacuo and the crude mixture was separated by silica gel chromatography (19:1; $CHCl_3$/acetone). Dicyclohexylurea (DCU, a by-product of the Steglich reaction) was removed by preparative HPLC.

Method B: Retinoic acid (**2**, 0.3 mmol, 1.5 eq), flavonolignan (**3–5**, 0.2 mmol, 1 eq), N-ethyl-N'-(3-dimethylaminopropyl)carbodiimide hydrochloride (EDC·HCl, 0.4 mmol, 2 eq) and DMAP (catalytic amount) were dissolved in dry THF (9 mL) and N,N-dimethylformamide (DMF, 1 mL). After stirring for 5 h in darkness under argon, the reaction mixture was diluted with EtOAc and the organic layer was separated and washed with water (2 × 15 mL). The combined organic layers were dried (Na_2SO_4), filtered and the solvent was removed in vacuo. Purification was performed by preparative HPLC and the fractions containing the product were further purified by silica chromatography ($CHCl_3$ 100% to $CHCl_3$/acetone; 19:1).

The reactions were monitored by thin layer chromatography (TLC) on silica gel plates (9:2:0.1; $CHCl_3$/acetone/formic acid).

2.4.2. Synthesis of Conjugates

Silybin-7-O-retinoate (**6ab**): Natural silybin (**3ab**) and retinoic acid (**2**) were reacted according to the method A or method B to yield title compound **6ab** as yellow solid; method A (32 mg, 0.042 mmol, 13.2%); method B (22 mg, 0.028 mmol, 9.6%). For [^1]H and [^13]C NMR data see Table S1 in the Supplementary Material. HRESIMS *m/z*: [M − H]⁻ calcd for $C_{45}H_{47}O_{11}$ 763.31239; found 763.31182 (Figure S1).

Silybin A-7-O-retinoate (**6a**): Silybin A (**3a**) and retinoic acid (**2**) were reacted according to the method A or method B to yield title compound **6a** as yellow solid; method A (62 mg, 0.081 mmol, 25%); method B (23 mg, 0.030 mmol, 10%). For [^1]H and [^13]C NMR data see Table S2 in the Supplementary Material. HRESIMS *m/z*: [M − H]⁻ calcd for $C_{45}H_{47}O_{11}$ 763.31239; found 763.31127 (Figure S2).

Silybin B-7-O-retinoate (**6b**): Silybin B (**3b**) and retinoic acid (**2**) were reacted according to the method A or method B to yield title compound **6b** as yellow solid; method A (76 mg, 0.099 mmol, 31%); method B (44 mg, 0.058 mmol 19.2%). For [^1]H and [^13]C NMR data see Table S3 in the Supplementary Material. HRESIMS *m/z*: [M − H]⁻ calcd for $C_{45}H_{47}O_{11}$ 763.31239; found 763.31138 (Figure S3).

2,3-Dehydrosilybin AB-3-O-retinoate (**7**): Racemic 2,3-dehydrosilybin (**4**) and retinoic acid were reacted according to the method A or method B to yield title compound **7** as yellow solid; method A (52 mg, 0.068 mmol, 21%); method B (49 mg, 0.064 mmol, 21%). For [^1]H and [^13]C NMR data see Table S4 in the Supplementary Material. HRESIMS *m/z*: [M − H]⁻ calcd for $C_{45}H_{45}O_{11}$ 761.29674; found 761.29657 (Figure S4).

Isosilybin A-7-O-retinoate (**8a**): Isosilybin A (**5a**) and retinoic acid (**2**) were reacted according to the method A or method B to yield title compound **8a** as yellow solid; method A (58 mg, 0.076 mmol, 24%); method B (57 mg, 0.075 mmol, 25%). For [^1]H and [^13]C NMR data see Table S5 in the Supplementary Material. HRESIMS *m/z*: [M − H]⁻ calcd for $C_{45}H_{47}O_{11}$ 763.31239; found 763.31197 (Figure S5).

Isosilybin B-7-O-retinoate (**8b**): Isosilybin B (**5b**) and retinoic acid (**2**) were reacted according to the method A to yield title compound **8b** as yellow solid (23 mg, 0.030 mmol, 9.5%). For ^{1}H and ^{13}C NMR data see Table S6 in the Supplementary Material. HRESIMS *m/z*: [M − H]$^-$ calcd for $C_{45}H_{47}O_{11}$ 763.31239; found 763.31207 (Figure S6).

The purity of all compounds used for consecutive tests was over 91% as determined by HPLC.

2.5. Antioxidant Activity

2.5.1. Determination of Log P Values

The lipophilicity/hydrophilicity of the compounds (miLogP) was calculated using the Molinspiration property engine v2016.10 (http://www.molinspiration.com, Molinspiration Cheminformatics, Slovensky Grob, Slovakia) [17].

2.5.2. DPPH Radical Scavenging

The radical scavenging activity of the prepared conjugates, their parent compounds, their equimolar mixtures and retinol was evaluated as their capacity to scavenge 1,1-diphenyl-2-picrylhydrazyl radicals (DPPH, Sigma-Aldrich, Prague, Czech Republic) [18], as described previously [10] with minor modifications. Briefly, 15 µL of the tested substance (final concentration 0–20 mM in DMSO) was mixed with 285 µL of a freshly prepared methanolic DPPH solution (final concentration 20 µM) in a microtiter plate well. After 30 min at 25 °C, the absorbance at 517 nm was measured. The activity was expressed as the concentration of the compound required for reducing the absorbance to 50% (IC$_{50}$) of its initial value.

2.5.3. Antioxidant Activity

Ferric reducing antioxidant power (FRAP) [19] and cupric reducing antioxidant capacity (CUPRAC) [20] were measured using kits from Bioquochem (Llanera-Asturias, Spain) according to the manufacturer's instructions and expressed as trolox equivalents (TE). To determine oxygen radical absorption capacity (ORAC) [21], 73 µL of fluorescein (1.8 mg/L) solution in phosphate buffered saline (PBS, pH 7.4) was mixed with 2 µL of the samples (0–20 µM) in the wells of a 96-well plate. After 15 min incubation (37 °C), 25 µL of freshly prepared 2,2'-azo-*bis*(2-amidinopropane) dihydrochloride (AAPH, 60 mg/mL) was added to each well except for the negative control, where AAPH was replaced for PBS. The fluorescence (ex./em., 485/535 nm) was recorded immediately and for the next 2 h with the measurement step for 5 min using a microplate reader (SpectraMax i3 Multi-Mode Detection Platform, Molecular Devices, Wokingham, UK).

For cellular antioxidant activity (CAA) assay [22], the HepG2 cell line (ATCC, CCL-23TM, USA) was cultivated in Eagle's Minimum Essential Medium (Sigma-Aldrich, USA) supplemented with 10% of foetal bovine serum, 2 mM L-glutamine and 1× Antibiotic Antimycotic Solution (Sigma-Aldrich, St. Louis, MO, USA) in a CO_2 incubator (5% CO_2, 37 °C, Thermo Fisher Scientific, Waltham, MA, USA) and passaged 2×/week using trypsin-EDTA solution. For the experiments, 100 µL of the cells (1 × 10^6 cells/mL, Cellometer Auto T4 Bright Field Cell Counter, Nexcelom Bioscience, Lawrence, MA, USA) were split into 96-well plates. After 24 h, the cells were washed 3× with PBS (MultiFlo Multi-Mode Dispenser, BioTek, USA) and Dulbecco's Modified Eagle's Medium (Sigma-Aldrich, St. Louis, MO, USA) supplemented with the 2',7'-dichlorodihydrofluorescein diacetate (DCFH-DA, 0.0125 mg/mL) was added to each well together with the tested samples in the concentration range 0–500 µM. After incubation (1 h, CO_2 incubator, 37 °C), the medium was manually replaced for AAPH solution (0.16 mg/mL) and the fluorescence (ex./em. 485/540 nm) was recorded immediately for 2 h with 5 min steps.

2.5.4. Statistical Analysis

All data were analysed with one-way ANOVA, Scheffé and least squares difference tests for post hoc comparisons among pairs of means using the statistical package Statext ver. 2.1 (STATEXT LLC, Wayne, NJ, USA). Differences were considered statistically significant when $p < 0.05$.

3. Results and Discussion

3.1. Synthesis of Conjugates

Due to the inherent sensitivity of carotenoids and flavonoids to oxidation under harsh conditions in organic syntheses, an enzymatic approach using the lipase Novozym 435® with a cross linker was first tested for the linking of carotenoids with flavonoids. Such procedures have the advantage of being chemo-, regio- and in some cases even stereoselective while working in mild conditions [23] and Novozym 435® is highly selective for primary alcohol groups [10]. This lipase or various oxidases were previously used to prepare dimers of silybin (**3ab**) [24] and 2,3-dehydrosilybin (**4**), as well as heterodimers [25], combined with different flavonoids, which showed interesting antioxidant and biological activities. Also, retinol (**1**) was conjugated with dicarboxylic acids, lactate and oleate under lipase catalysis to reduce photodestruction and possible irritation when used in cosmetics [26]. By chemical synthesis, a set of dimers of various carotenoid-linked dicarboxylates was prepared [27]. Water-soluble carotenoid conjugates with hydrophilic polyols linked by dicarboxylic linkers were prepared using lipase [28]. An interesting heterodimer of retinol was prepared by chemically catalysed esterification of retinol (**1**) and retinoic acid (**2**) [29]. There are many examples of coupling retinol (**1**) and its derivatives to palmitic acid to make a long aliphatic ester [30].

However, the enzymatic method proved to be unfeasible for the conjugation of retinol with the flavonolignans **3ab**, **3a**, **3b**, **4**, **5a** and **5b** and some previously published enzymatic methods for retinol conjugation [28] failed to be reproduced in our hands. Therefore, an original, so far unexplored chemical approach for the ester linkage formation was chosen. The classic Steglich approach was utilized using DCC and a catalytic amount of DMAP (Scheme 1). The reactions proceeded to yield the derivatives with the retinoic moiety at the C-7 position of the flavonolignan except for 2,3-dehydrosilybin where the 3-O-retinyl-2,3-dehydrosilybin was isolated as the major product. During Steglich reaction a by-product DCU was formed, which could be removed using ASAHIPAK GS-310 20F column, a hitherto unknown procedure for DCU removal. Unfortunately, during the separation of the products the parent flavonolignan also co-eluted with the respective product. The product was then purified by silica chromatography in darkness. To overcome the contamination problem EDC·HCl was used, which could easily be removed by washing with water.

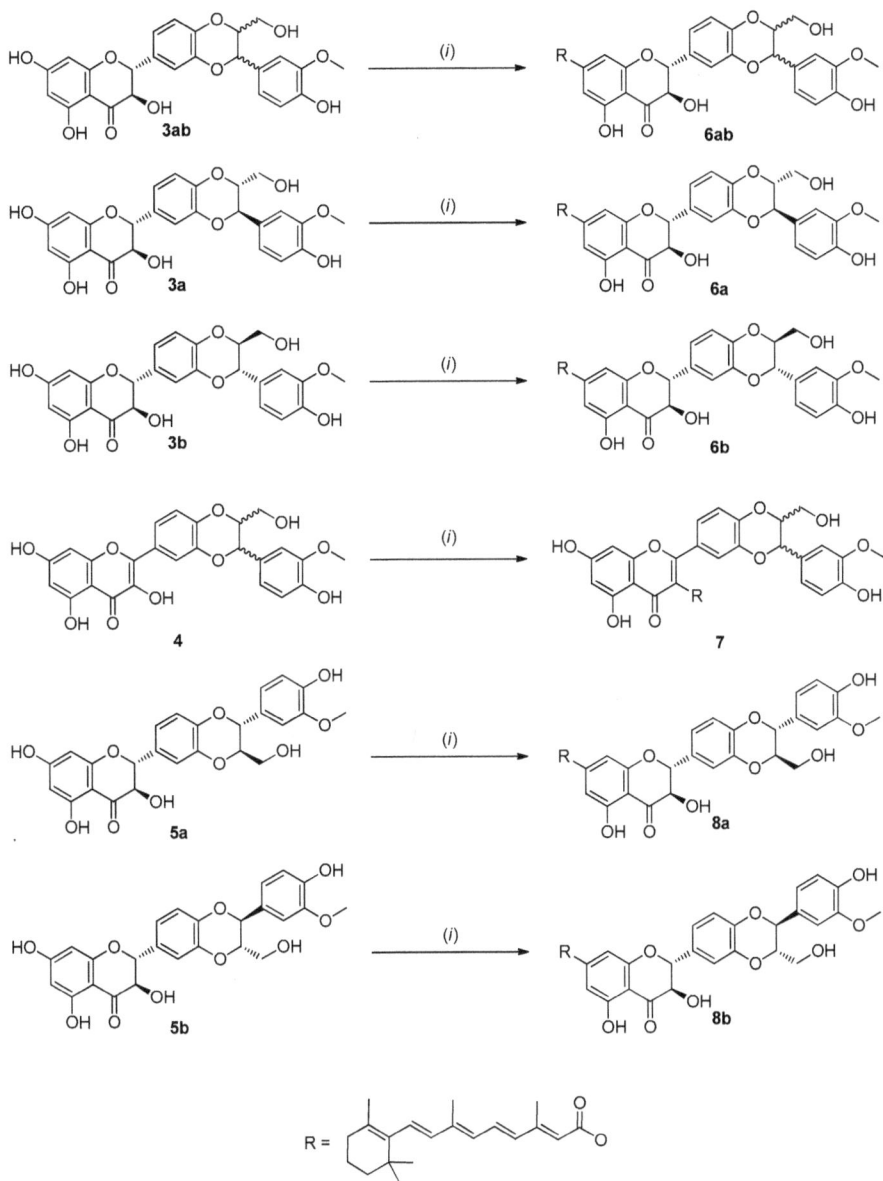

Scheme 1. Esterification of the flavonolignans with retinoic acid. (*i*) Method A: **2**, DCC, DMAP, THF, r.t., 18 h or Method B: **2**, EDC·HCl, DMAP, THF, 5 h.

All newly prepared carotenoyl-flavonolignans were fully characterized by ^{1}H and ^{13}C NMR and their structures were confirmed using HRMS. In general, the isolated yields are just moderate ranging from 9–25%, which was caused partly by decomposition of the starting material and often by formation of rather complex reaction mixtures, which were not quite easy to separate even using preparative HPLC. This esterification was regioselective providing nearly exclusively 7-O-esters of flavonoids, which is a great advantage as it avoids tedious protection/deprotection procedures of multifunctional (five OH groups) flavonolignan structure. Only in the case of 2,3-dehydrosilybin the 3-*O*-ester was preferentially formed probably due to higher reactivity of C-3 OH in 2,3-dehydroflavonolignans [10]. Our new method enabled to obtain sufficient amounts of heteroconjugates for antioxidant and further biophysical tests.

3.2. Antioxidant and Biophysical Testing of Conjugates

The oxidations in living organism caused by free radicals are usually undesirable processes damaging such important biomolecules as DNA, proteins and lipids [31]. The application of antioxidants is one of the most straightforward means to protect these biomolecules from oxidative stress. Many assays are available for determination of antioxidant potential in vitro. These assays are based on the reduction of stable radicals (DPPH), on the competitive bleaching of a probe (ORAC) or on the reduction of iron ions (CUPRAC, FRAP) [32]. CUPRAC test is very similar to FRAP assay but is based on the reduction of Cu^{2+} in a neutral pH, while FRAP test is based on Fe^{3+} in acidic pH. As each assay suffers from some limitations, a set of antioxidant tests was used here to combine more parameters.

DPPH test is one of the most commonly performed antioxidant assays with natural and (semi)synthetic biologically active compounds. Although it has no direct physiological relevance, this assay allows quick comparison of free radical scavenging potential of new derivatives as this activity has been published for many compounds [33]. We have therefore tested our conjugates for DPPH scavenging and compared their activity with the activity of the parent compounds, that is, retinoic acid and the respective flavonolignan and their equimolar mixtures; retinol (**1**) was used as a benchmark. As expected, the activity of retinoic acid (**2**) was about half that of retinol (**1**) with IC_{50} values 745 and 1485 µM, respectively. In accordance with our previously published data [25,34–36] also parent flavonolignans **3ab**, **3a**, **3b** and **5a** displayed relatively low activity (IC_{50} 472–818 µM), while 2,3-dehydrosilybin (**4**) was much more active (19.2 µM, Table 1). The conjugates **6a** and **6b** with silybin diastereomers displayed significantly better DPPH scavenging activity (IC_{50} 379 and 540 µM) than both the parent compounds and their equimolar mixtures. In contrast, the conjugates **6ab**, **7** and **8a** with natural silybin, 2,3-dehydrosilybin and isosilybin A were significantly worse scavengers than the parent compounds and their equimolar mixtures (Table 1). We could only speculate on the reasons why diastereomerically pure conjugates have better antioxidant activity than their equimolar mixture. Nevertheless, even though the differences are significant, the IC_{50} values of all silybin isomers and conjugates are in the same order of magnitude and similar to that of retinol and isosilybin, which all are relatively weak antioxidants. The biological relevance of such differences is probably rather low.

Table 1. 1,1-Diphenyl-2-picrylhydrazyl (DPPH) radical scavenging activities (IC_{50} [µM]) of retinoic conjugates in comparison with their parent compounds and their equimolar mixtures.

	Parent Compound	Conjugate with Retinoic Acid	Mixture 1:1 Eq with Retinoic Acid
Retinol (**1**)	745 ± 11	–	–
Retinoic acid (**2**)	1485 ± 110	–	–
Silybin (**3ab**)	472 ± 16	(**6ab**) 666 ± 16 *	499 ± 6
Silybin A (**3a**)	818 ± 22	(**6a**) 379 ± 19 *	750 ± 33
Silybin B (**3b**)	659 ± 29	(**6b**) 540 ± 24 *	773 ± 7
2,3-Dehydrosilybin (**4**)	19.2 ± 0.3	(**7**) 734 ± 35 *	15.1 ± 0.3
Isosilybin A (**5a**)	783 ± 9	(**8a**) 2361 ± 152 *	610 ± 18

Results are presented as means ± standard error from at least three independent experiments. * Values are significantly different ($p < 0.05$) compared to the parent flavonolignan, retinoic acid and their equimolar mixture.

The main disadvantage of DPPH scavenging assay is the use of an unnatural radical [37]. Such methods based on competitive probe reactions or indirect methods based on persistent radicals should only be used for preliminary screening purposes [37]. Thus, to better characterize the potential of the obtained conjugates in terms of their biological activity, a series of simple, rapid, sensitive and reproducible biochemical antioxidant assays including FRAP, CUPRAC and ORAC [38] and a more relevant cellular antioxidant activity (CAA [37]) was measured (Table 2). Many samples having the ability to reduce radicals in chemical assays were shown to fail in cellular assays, which also take into account the bioavailability and first-pass metabolism of tested compounds [39]. Therefore, in this study, we compared two methods with the same mechanism of action: ORAC assay was performed as a classical biochemical method presenting the ability of tested compounds to quench peroxyl radicals generated from AAPH to protect fluorescein from oxidation. This assay serves better as a physical description of the tested compounds [37]. To reflect biological aspect, CAA was measured using the liver carcinoma HepG2 cell line where the compound must enter the cell to fulfil its role as an antioxidant. The CAA assay partially includes the bioavailability, aspects of uptake and metabolism. Despite the mentioned advantages, the CAA does not provide the insight into the fate of tested compounds in the whole organism including distribution, clearance and for example, the ability of tested compounds to induce the transcription of antioxidant enzymes [37].

Table 2. Antioxidant activities and lipophilicity/hydrophilicity of retinoic conjugates in comparison with their parent compounds.

	FRAP [TE] [a]	CUPRAC [b] [TE]	ORAC [c] (IC_{50} [μM])	CAA [d] (IC_{50} [μM])	LogP [e]
Retinol (1)	1.76 ± 0.04	0.09 ± 0.01[f]	169 ± 12	1271 ± 147	5.92
Retinoic acid (2)	0.62 ± 0.02	0.10 ± 0.02[f]	13 ± 2	460 ± 211	5.80
Silybin (3ab)	0.335 ± 0.006	0.17 ± 0.00	7.8 ± 0.7	11.8 ± 0.3	1.47
Silybin A (3a)	0.278 ± 0.005	0.20 ± 0.00	8.5 ± 0.4	10.0 ± 0.7	1.47
Silybin B (3b)	0.268 ± 0.009	0.16 ± 0.02	8.0 ± 0.3	6.8 ± 0.5	1.47
2,3-Dehydrosilybin (4)	4.06 ± 0.05	0.25 ± 0.00	9.6 ± 0.4	10.9 ± 0.5	2.44
Isosilybin A (5a)	0.280 ± 0.006	0.16 ± 0.02	4.3 ± 1.1	>100 [h]	1.47
Silybin AB-7-O-retinoate (6ab)	0.03 ± 0.02 [*,#]	0.22 ± 0.01[#]	91 ± 4 [*,#]	>500 [h]	7.53
Silybin A-7-O-retinoate (6a)	0.038 ± 0.002 [*,#]	0.21 ± 0.02[#]	9.0 ± 0.7 [j]	>50 [g,h]	7.53
Silybin B-7-O-retinoate (6b)	0.010 ± 0.001 [*,#]	0.04 ± 0.01[f]	230 ± 8 [*,#]	>500 [g]	7.53
2,3-Dehydrosilybin-3-O-retinoate (7)	0.034 ± 0.005 [*,#]	0.07 ± 0.02[f]	130 ± 7 [*,#]	>500 [g]	8.21
Isosilybin A-7-O-retinoate (8a)	0.022 ± 0.003 [*,#]	0.11 ± 0.01[f]	174 ± 7 [*,#]	>500 [g]	7.53

Results are presented as means ± standard error from at least three independent experiments. [a] Ferric-reducing antioxidant potential (trolox equivalents, TE); [b] cupric reducing antioxidant capacity; [c] oxygen radical absorption capacity, [d] cellular antioxidant activity; [e] hydrophobicity of compounds; [*,#] values significantly different ($p < 0.05$) compared to the parent flavonolignan (*) and retinoic acid (#); [f] measurement disturbed by the formation of a precipitate in the reaction mixture; [g] the sample decomposed; [h] no activity noted at the highest concentration tested.

In most of these tests all the conjugates displayed significantly lower activity than their respective parent compounds. In the CUPRAC assay intended specifically for hydrophobic antioxidants, where in some cases the conjugates **6ab** and **6b** had an activity comparable to the parent flavonolignans, most samples containing a carotenoid moiety precipitated in the reaction mixture thus hampering the activity determination and making the comparison mostly impossible. Dose-dependent curves were obtained in both ORAC and CAA assay for single compounds resulting in IC_{50} determination (Table 2). In contrast, no anti-oxidant activity of the conjugates was detected in liver cells in all the concentration range tested up to 50 or 500 μM. In fact, such a large concentration is not expectable in plasma. Furthermore, the conjugate **6a** proved to be unstable and decomposed during ORAC and CAA measurement.

Diastereomerically pure conjugates **6a** and **6b** displayed significantly better DPPH scavenging activity (IC_{50} 379 and 540 μM) than both the parent compounds and their equimolar mixtures. This finding could have potential exploitation as silybin is the amplest and largely used flavonoid from the silymarin complex. New, so far undescribed, conjugates of retinoic acid and flavonolignans have obviously comparable or even worse antioxidant activity in relation to their parent molecules. We could speculate that one reason could be the blocking of the highly (anti-radical) active [40] moiety at C-7 of flavonolignans and/or substantial increase of lipophilicity of the conjugates (logP 7.53 and

8.21, Table 2). Highly lipophilic conjugates are likely to be incorporated into the cell membrane and do not pass into the cells to exert their effect topically in the membrane.

4. Conclusions

We have described to our best knowledge the first synthesis of a series of hybrid ester conjugates of retinoic acid with various flavonolignans (flavonoid-carotenoid supramolecular conjugates) such as silybin, 2,3-dehydrosilybin and isosilybin. Antioxidant/anti-radical activities and bio-physical properties of novel carotenoid-flavonoid hybrids as well as various mixtures of the respective components were investigated. The conjugates with silybin diastereomers, which are the most important flavonolignans in silymarin complex, displayed better DPPH scavenging activity than both the parent compounds and their equimolar mixtures. Other conjugates have comparable or even worse antioxidant activity in relation to their parent molecules.

Supplementary Materials: Supplementary material containing MS, ^{1}H and ^{13}C NMR data of the new compounds can be found at http://www.mdpi.com/2076-3921/8/7/236/s1.

Author Contributions: Individual contributions of the authors: conceptualization, V.K. and K.V.; methodology, C.S.C., D.B., C.S.C. and K.V.; HPLC analysis, L.P.; antioxidant activity, K.V. and J.V.; NMR analysis, M.K.; MS analysis, J.C.; writing—original draft preparation, C.S.C.; writing—review and editing, all authors; supervision, V.K.; funding acquisition, V.K.

Funding: This research was funded by Ministry of Education, Youth and Sports of the Czech Republic, grant number LTC17009 and by ESF COST Action CA15136 "EUROCAROTEN".

Acknowledgments: The authors acknowledge Josef Cvačka from the Institute of Organic Chemistry and Biochemistry of the Czech Academy of Sciences, Prague, Czech Republic for HRMS measurements.

Conflicts of Interest: The authors declare no conflict of interest.

Abbreviations

AAPH	2,2′-Azo-*bis*(2-amidinopropane) dihydrochloride
BHT	Butylated hydroxytoluene
CAA	Cellular antioxidant activity
CUPRAC	Cupric reducing antioxidant capacity
DCC	N,N′-Dicyclohexylcarbodiimide
DCFH-DA	2′,7′-Dichlorodihydrofluorescein diacetate
DCU	Dicyclohexylurea
DMAP	4-Dimethylaminopyridine
DPPH	1,1-Diphenyl-2-picrylhydrazyl radical
EDC	1-Ethyl-3-(3-dimethylaminopropyl)carbodiimide
FRAP	Ferric reducing antioxidant power
NF-κb	Nuclear factor kappa-light-chain-enhancer of activated B cells
ORAC	Oxygen radical absorbance capacity
ROS	Reactive oxygen species
THF	Tetrahydrofuran

References

1. Brambilla, D.; Mancuso, C.; Scuderi, M.R.; Bosco, P.; Cantarella, G.; Lempereur, L.; Di Benedetto, G.; Pezzino, S.; Bernardini, R. The role of antioxidant supplement in immune system, neoplastic and neurodegenerative disorders: A point of view for an assessment of the risk/benefit profile. *Nutrition* **2008**, *7*, 29. [CrossRef] [PubMed]
2. Stahl, W.; Sies, H. Carotenoids and flavonoids contribute to nutritional protection against skin damage from sunlight. *Mol. Biotechnol.* **2007**, *37*, 26–30. [CrossRef] [PubMed]

3. Beutner, S.; Bloedorn, B.; Frixel, S.; Hernández Blanco, I.; Hoffmann, T.; Martin, H.-D.; Mayer, B.; Noack, P.; Ruck, C.; Schmidt, M.; et al. Quantitative assessment of antioxidant properties of natural colorants and phytochemicals: Carotenoids, flavonoids, phenols and indigoids. The role of β-carotene in antioxidant functions. *J. Sci. Food Agric.* **2001**, *81*, 559–568. [CrossRef]

4. Forman, H.J.; Davies, K.J.; Ursini, F. How do nutritional antioxidants really work: Nucleophilic tone and para-hormesis versus free radical scavenging in vivo. *Free Radic. Biol. Med.* **2014**, *66*, 24–35. [CrossRef] [PubMed]

5. Paiva, S.A.R.; Russell, R.M. β-Carotene and other carotenoids as antioxidants. *J. Am. Coll. Nutr.* **1999**, *18*, 426–433. [CrossRef] [PubMed]

6. Pietta, P.G. Flavonoids as antioxidants. *J. Nat. Prod.* **2000**, *63*, 1035–1042. [CrossRef] [PubMed]

7. Koufaki, M.; Calogeropoulou, T.; Rekka, E.; Chryselis, M.; Papazafiri, P.; Gaitanaki, C.; Makriyannis, A. Bifunctional agents for reperfusion arrhythmias: Novel hybrid vitamin E/Class I antiarrhythmics. *Bioorg. Med. Chem.* **2003**, *11*, 5209–5219. [CrossRef] [PubMed]

8. Bebbington, D.; Dawson, C.E.; Gaur, S.; Spencer, J. Prodrug and covalent linker strategies for the solubilization of dual-action antioxidants/iron chelators. *Bioorg. Med. Chem. Lett.* **2002**, *12*, 3297–3300. [CrossRef]

9. Gažák, R.; Marhol, P.; Purchartová, K.; Monti, D.; Biedermann, D.; Riva, S.; Cvak, L.; Křen, V. Large-scale separation of silybin diastereoisomers using lipases. *Process Biochem.* **2010**, *45*, 1657–1663. [CrossRef]

10. Vavříková, E.; Křen, V.; Ježová-Kalachová, L.; Biler, M.; Chantemargue, B.; Pyszková, M.; Riva, S.; Kuzma, M.; Valentová, K.; Ulrichová, J.; et al. Novel flavonolignan hybrid antioxidants: From enzymatic preparation to molecular rationalization. *Eur. J. Med. Chem.* **2017**, *127*, 263–274. [CrossRef]

11. Papa, T.B.R.; Pinho, V.D.; do Nascimento, E.S.P.; Santos, W.G.; Burtoloso, A.C.B.; Skibsted, L.H.; Cardoso, D.R. Astaxanthin diferulate as a bifunctional antioxidant. *Free Radic. Res.* **2015**, *49*, 102–111. [CrossRef] [PubMed]

12. Hundsdörfer, C.; Stahl, W.; Müller, T.J.J.; De Spirt, S. UVA photoprotective properties of an artificial carotenylflavonoid hybrid molecule. *Chem. Res. Toxicol.* **2012**, *25*, 1692–1698. [CrossRef] [PubMed]

13. Beutner, S.; Frixel, S.; Ernst, H.; Hoffmann, T.; Hernandez-Blanco, I.; Hundsdoerfer, C.; Kiesendahl, N.; Kock, S.; Martin, H.D.; Mayer, B.; et al. Carotenylflavonoids, a novel group of potent, dual-functional antioxidants. *Arkivoc* **2007**, *8*, 279–295.

14. Han, R.M.; Zhang, J.P.; Skibsted, L.H. Reaction dynamics of flavonoids and carotenoids as antioxidants. *Molecules* **2012**, *17*, 2140–2160. [CrossRef] [PubMed]

15. Gažák, R.; Fuksová, K.; Marhol, P.; Kuzma, M.; Agarwal, R.; Křen, V. Preparative method for isosilybin isolation based on enzymatic kinetic resolution of silymarin mixture. *Process Biochem.* **2013**, *48*, 184–189. [CrossRef]

16. Džubák, P.; Hajdúch, M.; Gažák, R.; Svobodová, A.; Psotová, J.; Walterová, D.; Sedmera, P.; Křen, V. New derivatives of silybin and 2,3-dehydrosilybin and their cytotoxic and p-glycoprotein modulatory activity. *Bioorg. Med. Chem.* **2006**, *14*, 3793–3810. [CrossRef] [PubMed]

17. Pyka, A.; Babuska, M.; Zachariasz, M. A comparison of theoretical methods of calculation of partition coefficients for selected drugs. *Acta Pol. Pharm.* **2006**, *63*, 159–167. [PubMed]

18. Joyeux, M.; Mortier, F.; Fleurentin, J. Screening of antiradical, antilipoperoxidant and hepatoprotective effect of nine plant extracts used in Caribbean folk medicine. *Phythother. Res.* **1995**, *9*, 228–230. [CrossRef]

19. Jones, A.; Pravadali-Cekic, S.; Dennis, G.R.; Bashir, R.; Mahon, P.J.; Shalliker, R.A. Ferric reducing antioxidant potential (FRAP) of antioxidants using reaction flow chromatography. *Anal. Chim. Acta* **2017**, *967*, 93–101. [CrossRef]

20. Özyürek, M.; Güçlü, K.; Apak, R. The main and modified CUPRAC methods of antioxidant measurement. *Trends Anal. Chem.* **2011**, *30*, 652–664. [CrossRef]

21. Huang, D.; Ou, B.; Hampsch-Woodill, M.; Flanagan, J.A.; Prior, R.L. High-throughput assay of oxygen radical absorbance capacity (ORAC) using a multichannel liquid handling system coupled with a microplate fluorescence reader in 96-well format. *J. Agric. Food Chem.* **2002**, *50*, 4437–4444. [CrossRef] [PubMed]

22. Wolfe, K.L.; Liu, R.H. Cellular antioxidant activity (CAA) assay for assessing antioxidants, foods and dietary supplements. *J. Agric. Food Chem.* **2007**, *55*, 8896–8907. [CrossRef] [PubMed]

23. Antonopoulou, I.; Varriale, S.; Topakas, E.; Rova, U.; Christakopoulos, P.; Faraco, V. Enzymatic synthesis of bioactive compounds with high potential for cosmeceutical application. *Appl. Microbiol. Biotechnol.* **2016**, *100*, 6519–6543. [CrossRef] [PubMed]

24. Gažák, R.; Sedmera, P.; Marzorati, M.; Riva, S.; Křen, V. Laccase-mediated dimerization of the flavonolignan silybin. *J. Mol. Catal. B* **2008**, *50*, 87–92. [CrossRef]

25. Vavříková, E.; Vacek, J.; Valentová, K.; Marhol, P.; Ulrichová, J.; Kuzma, M.; Křen, V. Chemo-enzymatic synthesis of silybin and 2,3-dehydrosilybin dimers. *Molecules* **2014**, *19*, 4115–4134. [CrossRef] [PubMed]

26. Maugard, T.; Legoy, M.D. Enzymatic synthesis of derivatives of vitamin a in organic media. *J. Mol. Catal. B Enzym.* **2000**, *8*, 275–280. [CrossRef]

27. Háda, M.; Nagy, V.; Takátsy, A.; Deli, J.; Agócs, A. Dicarotenoid esters of bivalent acids. *Tetrahedron Lett.* **2008**, *49*, 3524–3526. [CrossRef]

28. Rejasse, B.; Maugard, T.; Legoy, M.D. Enzymatic procedures for the synthesis of water-soluble retinol derivatives in organic media. *Enzyme Microb. Technol.* **2003**, *32*, 312–320. [CrossRef]

29. Kim, H.; Kim, B.; Kim, H.; Um, S.; Lee, J.; Ryoo, H.; Jung, H. Synthesis and in vitro biological activity of retinyl retinoate, a novel hybrid retinoid derivative. *Bioorg. Med. Chem.* **2008**, *16*, 6387–6393. [CrossRef]

30. Liu, Z.Q.; Zhou, L.M.; Liu, P.; Baker, P.J.; Liu, S.S.; Xue, Y.P.; Xu, M.; Zheng, Y.G. Efficient two-step chemo-enzymatic synthesis of all-trans-retinyl palmitate with high substrate concentration and product yield. *Appl. Microbiol. Biotechnol.* **2015**, *99*, 8891–8902. [CrossRef]

31. Akono Ntonga, P.; Baldovini, N.; Mouray, E.; Mambu, L.; Belong, P.; Grellier, P. Activity of *Ocimum basilicum*, *Ocimum canum* and *Cymbopogon citratus* essential oils against *Plasmodium falciparum* and mature-stage larvae of *Anopheles funestus* s.s. *Parasite* **2014**, *21*, 33. [CrossRef] [PubMed]

32. Chauhan, N.; Malik, A.; Sharma, S.; Dhiman, R.C. Larvicidal potential of essential oils against *Musca domestica* and *Anopheles stephensi*. *Parasitol. Res.* **2016**, *115*, 2223–2231. [CrossRef] [PubMed]

33. Huang, D.; Ou, B.; Prior, R.L. The chemistry behind antioxidant capacity assays. *J. Agric. Food Chem.* **2005**, *23*, 1841–1856. [CrossRef] [PubMed]

34. Valentová, K.; Biedermann, D.; Křen, V. 2,3-Dehydroderivatives of silymarin flavonolignans: Prospective natural compounds for the prevention of chronic diseases. *Proceedings* **2019**, *11*, 21. [CrossRef]

35. Valentová, K.; Purchartová, K.; Rydlová, L.; Roubalová, L.; Biedermann, D.; Petrásková, L.; Křenková, A.; Pelantová, H.; Holečková-Moravcová, V.; Tesařová, E.; et al. Sulfated metabolites of flavonolignans and 2,3-dehydroflavonolignans: Preparation and properties. *Int. J. Mol. Sci.* **2018**, *19*, 2349. [CrossRef] [PubMed]

36. Pyszková, M.; Biler, M.; Biedermann, D.; Valentová, K.; Kuzma, M.; Vrba, J.; Ulrichová, J.; Sokolová, R.; Mojovic, M.; Popovic-Bijelic, A.; et al. Flavonolignan 2,3-dehydroderivatives: Preparation, antiradical and cytoprotective activity. *Free Radic. Biol. Med.* **2016**, *90*, 114–125. [CrossRef]

37. Amorati, R.; Valgimigli, L. Advantages and limitations of common testing methods for antioxidants. *Free Radic. Res.* **2015**, *49*, 633–649. [CrossRef]

38. Thaipong, K.; Boonprakob, U.; Crosby, K.; Cisneros-Zevallos, L.; Hawkins Byrne, D. Comparison of ABTS, DPPH, FRAP and ORAC assays for estimating antioxidant activity from guava fruit extracts. *J. Food Compos. Anal.* **2006**, *19*, 669–675. [CrossRef]

39. Blasa, M.; Angelino, D.; Gennari, L.; Ninfali, P. The cellular antioxidant activity in red blood cells (CAA-RBC): A new approach to bioavailability and synergy of phytochemicals and botanical extracts. *Food Chem.* **2011**, *125*, 685–691. [CrossRef]

40. Gažák, R.; Sedmera, P.; Vrbacký, M.; Vostálová, J.; Drahota, Z.; Marhol, P.; Walterová, D.; Křen, V. Molecular mechanisms of silybin and 2,3-dehydrosilybin antiradical activity—Role of individual hydroxyl groups. *Free Radic. Biol. Med.* **2009**, *46*, 745–758. [CrossRef]

antioxidants

MDPI

Article

Fucoxanthin—An Antibacterial Carotenoid

Tomasz M. Karpiński [1,*] and Artur Adamczak [2]

[1] Department of Medical Microbiology, Poznań University of Medical Sciences, Wieniawskiego 3, 61-712 Poznań, Poland
[2] Department of Botany, Breeding and Agricultural Technology of Medicinal Plants, Institute of Natural Fibres and Medicinal Plants, Kolejowa 2, 62-064 Plewiska, Poland
* Correspondence: tkarpin@ump.edu.pl or tkarpin@interia.pl; Tel.: +48-61-854-61-38

Received: 23 June 2019; Accepted: 22 July 2019; Published: 24 July 2019

Abstract: Fucoxanthin is a carotenoid produced by brown algae and diatoms. This compound has several biological properties such as antioxidant, anti-obesity, anti-diabetic, anticancer, and antimicrobial activities. Unfortunately, until now the latter effect has been poorly confirmed. The aim of this study was an evaluation of fucoxanthin activity against 20 bacterial species. Antimicrobial effect of fucoxanthin was determined by using the agar disc-diffusion and micro-dilution methods. The studied carotenoid acted against 13 bacteria growing in aerobic conditions. It was observed to have a significantly stronger impact on Gram-positive than Gram-negative bacteria. Mean zones of growth inhibition (ZOIs) for Gram-positive bacteria ranged between 9.0 and 12.2 mm, while for Gram-negative were from 7.2 to 10.2 mm. According to the agar disc-diffusion method, the highest activity of fucoxanthin was exhibited against *Streptococcus agalactiae* (mean ZOI 12.2 mm), *Staphylococcus epidermidis* (mean ZOI 11.2 mm), and *Staphylococcus aureus* (mean ZOI 11.0 mm), and in the microdilution test towards *Streptococcus agalactiae* with the minimal inhibitory concentration (MIC) of 62.5 μg/mL. On the other hand, fucoxanthin was not active against strict anaerobic bacteria.

Keywords: brown seaweeds; algal pigments; antibacterial activity; disk-diffusion method; micro-dilution method

1. Introduction

Edible seaweeds (red, brown and green marine macroalgae) are the rich source of various bioactive compounds: soluble dietary fibers, sulphated polysaccharides, phlorotannins, peptides, sulfolipids, polyunsaturated fatty acids, carotenoids, vitamins, and minerals [1]. Seaweed carotenoids mainly include: β-carotene, zeaxanthin, violaxanthin, lutein, and fucoxanthin [2]. Fucoxanthin is an orange-colored pigment predominantly found in brown algae (Phaeophyceae) and diatoms (Bacillariophyceae) [3,4]. It belongs to the class of xanthophylls and non-provitamin A carotenoids [5]. Its presence was confirmed, among others, in *Alaria crassifolia*, *Ascophyllum nodosum*, *Chaetoseros* sp., *Cladosiphon okamuranus*, *Cylindrotheca closterium*, *Cystoseira hakodatensis*, *Ecklonia stolonifera*, *Eisenia bicyclis*, *Fucus serratus*, *F. vesiculosus*, *Hijikia fusiformis*, *Himanthalia elongata*, *Ishige okamurae*, *Kjellmaniella crassifolia*, *Laminaria digitata*, *L. japonica*, *L. ochotensis*, *L. religiosa*, *Myagropsis myagroides*, *Odontella aurita*, *Padina tetrastromatica*, *Petalonia binghamiae*, *Phaeodactylum tricornutum*, *Sargassum fulvellum*, *S. heterophyllum*, *S. horneri*, *S. siliquastrum*, *Scytosiphon lomentaria*, *Sphaerotrichia divaricata*, *Turbinaria triquetra*, and *Undaria pinnatifida* [3–14]. Chemically, fucoxanthin is 3′-acetoxy-5,6-epoxy-3,5′-dihydroxy-6′,7′-didehydro-5,6,7,8,5′,6′-hexahydro-β,β-carotene-8-one with a molecular formula of $C_{42}H_{58}O_6$ and a molecular weight of 658.906 g/mol [15]. This metabolite includes a unique allenic bond and some oxygenic functional moieties such as epoxide, hydroxyl, carbonyl, and carboxyl groups [3]. The chemical structure of fucoxanthin is shown in Figure 1.

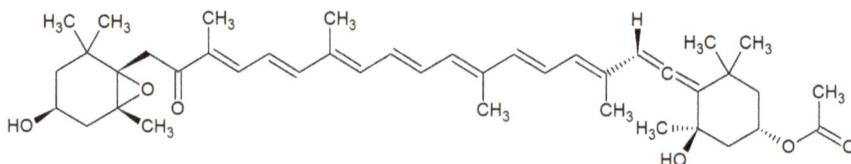

Figure 1. Chemical structure of fucoxanthin.

It has recently been shown that fucoxanthin exhibits a lot of biological properties, including a protective activity against oxidative stress. It was presented that fucoxanthin prevents the cytotoxic effect of the oxidative agent in a dose-dependent manner and had a protective effect against UV-B radiation and DNA damaging factors [10,16–19]. This compound additionally affected the lipid metabolism and possesses anti-obesity and anti-diabetic activities [11,20,21]. The influence of fucoxanthin on the weight loss, insulin resistance, and the lowering of the blood glucose level were confirmed [22]. It also had a beneficial impact on the cardiovascular system, manifested by a decrease in cholesterol and triacylglycerol levels, the lowering of the blood pressure, and the reduction of inflammatory processes [23].

Additionally, fucoxanthin demonstrated broad anticancer activity. The antiproliferative effect was stated in vitro among others against the following cell lines: leukemic (HD-60) [24,25], epithelial colorectal adenocarcinoma (Caco-2, DLD-1 and HT-29) [26], prostate cancer (PC-3, DU-145, LNCaP) [27–29], urinary bladder cancer (EJ-1) [30], osteosarcoma (Saos-2, MNNG/HOS and 143B) [31], breast cancer (MDA-MB-231) [32], non-small-cell lung cancer (NSCLC) [33], and gastric adenocarcinoma (MGC-803) [34]. Other studies presented that fucoxanthin acted preventively against cancer exerting the antiangiogenic, antilymphangiogenic, and antimetastatic effects [32,35,36].

In the literature, there is mentioned about antibacterial properties of fucoxanthin [37,38]. Screening of the PubMed/MEDLINE database carried out at the beginning of June 2019 with the search term "fucoxanthin" found more than 600 items, however the combination of the keywords "fucoxanthin" and "antibacterial" gave only nine records. Unfortunately, none of these articles concerns the antibacterial activity of fucoxanthin. In the electronic databases, we found only three publications presenting studies in this field [7,8,12].

Due to the lack of research and not fully confirmed antibacterial properties of fucoxanthin, the aim of the present study was an evaluation of this carotenoid activity against selected clinical strains of bacteria.

2. Materials and Methods

2.1. Microbial Strains and Culture Media

In this study, clinical strains of bacteria were used. None of the chosen strains was multi-resistant. Fucoxanthin was purchased from Sigma-Aldrich, Poland (product number: F6932, purity: ≥95%). Antimicrobial activity of this algal pigment was investigated against six Gram-positive bacteria (*Enterococcus faecalis*, *Staphylococcus aureus*, *S. epidermidis*, *Streptococcus agalactiae*, *S. pneumoniae*, and *S. pyogenes*), and seven Gram-negative bacteria (*Acinetobacter lwoffii*, *Escherichia coli*, *Klebsiella oxytoca*, *K. pneumoniae*, *Proteus mirabilis*, *Pseudomonas aeruginosa*, and *Serratia marcescens*) growing in aerobic conditions. The species of bacteria were grown at 35 °C for 24 h, in tryptone soy agar (TSA; Graso, Poland). Additionally, there were tested seven strict anaerobic bacteria (*Actinomyces israelii*, *Atopobium parvulum*, *Mitsuokella multacida*, *Peptococcus niger*, *Porphyromonas gingivalis*, *Propionibacterium acnes*, and *Veilonella parvula*). These species were cultured in anaerobic conditions using Genbox and Genbag anaer (bioMerieux, Poland) at 35 °C for 2–5 days, in Schaedler agar with 5% sheep blood (Graso, Poland). Two strains were tested for each species, except *A. israelii*, *M. multacida*, and *P. gingivalis* which were examined in one strain.

2.2. Antimicrobial Activity

The microbial growth inhibitory potential of the tested xanthophyll was determined by using the agar disc-diffusion method according to recommendations of the Clinical and Laboratory Standards Institute (CLSI) [39], and as described in our previous publication [40]. In brief, bacterial inocula of 0.5 McFarland were prepared. Next, 100 µL of all bacterial suspensions were inoculated on Mueller–Hinton agar with 5% sheep blood or Mueller–Hinton agar (Oxoid, Poland; Graso, Poland). Fucoxanthin was dissolved in 20% water solution of DMSO (Sigma-Aldrich, Poznań, Poland) in a final concentration of 1 mg/mL. A total of 25 µL of 1 mg/mL fucoxanthin (25 µg/disc) were transferred onto sterile filter papers (6 mm diameter). Additionally, sterile filter papers soaked 25 µL of 20% DMSO (negative control) were used. The plates were incubated at 35 °C for 18 h and anaerobes for two days. Results were shown as zones of growth inhibition (ZOIs). The experiments were repeated three times.

Minimal inhibitory concentration (MIC) was determined by the micro-dilution method using a 96-well plate (Nunc) according to CLSI [39]. Primarily, 100 µL of Mueller–Hinton broth, or Thioglicolate broth (Oxoid, Poland; Graso, Poland) for anaerobes, was placed in each well. The stock solution of fucoxanthin was transferred into the first well, and serial dilutions were performed so that concentrations in the range of 15.6 to 1000 µg/mL were obtained. The inoculums were adjusted to contain approximately 10^7 CFU/mL bacteria. 10 µL of the proper inoculums were added to the wells. Additionally, 10 µL of 0.2% aqueous solution of 2,3,5-triphenyltetrazolium chloride (TTC) was added to each well. TTC is converted in bacterial cells into red, insoluble formazan crystals [41]. Next, the plates were incubated at 35 °C for 24 h. The MIC value was taken as the lowest concentration of the extract that inhibited any visible bacterial growth. The experiments were repeated three times.

2.3. Statistical Analysis

The results reported in Table 1 are means ± SD of three parallel measurements, and medians. Data were tested using Statistica for Windows software. Statistical analysis of the results was based on Mann-Whitney U-test. Differences of $p < 0.05$ were considered to be significant.

3. Results

In the present research, the activity of fucoxanthin against bacterial strains belonging to 20 species was tested. This compound acted against 13 bacteria growing in aerobic conditions. It was observed a clearly stronger effect against Gram-positive than Gram-negative bacteria. Mean zones of growth inhibition (ZOIs) for Gram-positive bacteria were between 9.0 and 12.2 mm, while for Gram-negative ranged from 7.2 to 10.2 mm (Figure 2). The differences between the ZOIs of both groups were statistically significant ($p < 0.0001$). Minimal inhibitory concentrations (MICs) for Gram-positive bacteria reached values between 62.5 and 250 µg/mL (median 125 µg/mL), while for Gram-negative values were from 125 to 500 µg/mL (median 250 µg/mL). Between the MICs of both groups there were statistically significant differences ($p = 0.0009$). The highest activity of fucoxanthin in the agar disk-diffusion method was against *Streptococcus agalactiae* (mean ZOI 12.2 mm), *Staphylococcus epidermidis* (mean ZOI 11.2 mm), and *S. aureus* (mean ZOI 11.0 mm), and in the microdilution test towards *Streptococcus agalactiae* (MIC 62.5 µg/mL). Simultaneously, it was not found fucoxanthin's activity against seven strict anaerobic bacteria. The obtained values of the ZOIs and MICs are presented in Table 1.

Table 1. Antibacterial activity of fucoxanthin determined by the agar disc-diffusion and micro-dilution methods.

Studied Bacterial Strain	Zone of Growth Inhibition (ZOI) (mm)	Minimal Inhibitory Concentration (MIC) (µg/mL)
Gram-positive		
Enterococcus faecalis	9.0 ± 0.89	125–250
Staphylococcus aureus	11.0 ± 0.63	125
Staphylococcus epidermidis	11.2 ± 0.75	125
Streptococcus agalactiae	12.2 ± 0.75	62.5
Streptococcus pneumoniae	9.7 ± 0.52	125
Streptococcus pyogenes	10.0 ± 0.63	125
Mean of all ZOIs	10.5 ± 1.25	-
Median	10.0	125
Gram-negative		
Acinetobacter lwoffii	8.2 ± 0.41	250
Escherichia coli	10.2 ± 0.75	125
Klebsiella oxytoca	9.2 ± 0.75	125–250
Klebsiella pneumoniae	8.8 ± 0.75	250
Proteus mirabilis	7.2 ± 0.41	500
Pseudomonas aeruginosa	7.5 ± 0.55	250–500
Serratia marcescens	7.3 ± 0.52	500
Mean of all ZOIs	8.3 ± 1.18	-
Median	8.0	250
Anaerobic		
Actinomyces israelii	6.0	>1000
Atopobium parvulum	6.0	>1000
Mitsuokella multacida	6.0	>1000
Peptococcus niger	6.0	>1000
Porphyromonas gingivalis	6.0	>1000
Propionibacterium acnes	6.0	>1000
Veilonella parvula	6.0	>1000
Negative control		
20% DMSO	6.00 ± 0.00	-

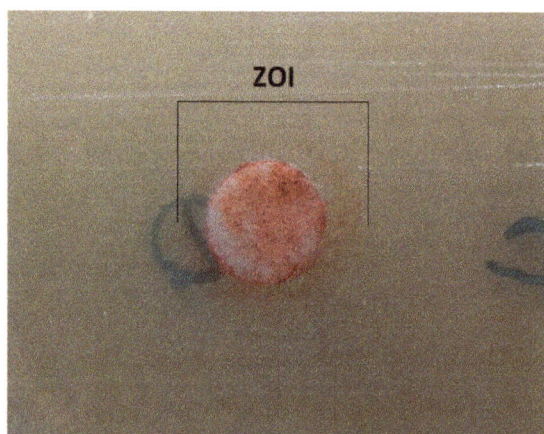

Figure 2. Culture of *Escherichia coli* strain. It is a visible zone of growth inhibition (ZOI) around the disk with 25 µg of fucoxanthin.

4. Discussion

In this paper, we presented the activity of fucoxanthin against 13 aerobic and 7 anaerobic bacteria. To our knowledge, this is the first work in which such a large number of species has been studied. Moreover, this is the first research of fucoxanthin, in which the minimal inhibitory concentrations (MICs) were tested. Fucoxanthin was investigated according to the disk-diffusion method at the concentration of 1 mg/mL. Additionally, it was used the microdilution assay in the levels of fucoxanthin from 15.6 to 1000 µg/mL. The results obtained in the second method amounted between 62.5 and 500 µg/mL. Deyab and Abou-Dobara [7] used this carotenoid isolated from the brown seaweed *Turbinaria triquetra* at the concentrations from 10 to 100 µg/mL. It can therefore be assumed that the used concentrations were similar. However, in our research the levels of the active compound were slightly higher. Unfortunately, in the above-mentioned article, the methodology of antimicrobial activity screening was not described, and the results were presented not according to the microbiological CLSI standards. These authors tested the activity of fucoxanthin against *E. coli, Bacillus cereus, B. subtilis, K. pneumoniae, S. aureus,* and *P. aeruginosa.* The zones of growth inhibitions for the above bacteria were 0.5–1.8 mm in the concentration of 10 µg/mL and 4.0–7.0 mm in 100 µg/mL. The ZOIs were very low and probably contrary to the guidelines for the disk-diffusion method, the width of the paper disk was not taken into account [7]. In our study, mean ZOIs amounted to 6.0 mm for anaerobic bacteria and from 7.2 to 12.2 mm for other species.

Rajauria and Abu-Ghannam [8] showed antimicrobial activity determined against *Listeria monocytogenes* using disc-diffusion method. The diameter of the growth inhibition zone reached 10.89 mm. The antimicrobial activity was demonstrated by both the purified fucoxanthin extracted from the brown alga *Himanthalia elongata* and chemical standard, at a concentration of 1 mg/mL (25 µg/disc). In our study, the same fucoxanthin concentration was used, and the ZOIs for tested bacteria were similar to this presented for *L. monocytogenes.*

Recently, the antibacterial properties of fucoxanthin were reported by Liu et al. [12]. This pigment was extracted from the edible seaweed *Undaria pinnatifida* with a purity of 82.70%, and tested against five human pathogens. According to the agar well diffusion method, fucoxanthin strongly inhibited the growth of Gram-positive bacteria: *B. subtilis, E. faecalis, S. aureus,* and *Enterococcus* sp. The diameters of their inhibition zones reached 25.49, 25.24, 21.80, and 12.66 mm, respectively. Similar to our results, they indicated weaker activity towards Gram-negative bacteria. In the case of *P. aeruginosa* it was 9.50 mm.

In the case of infection, mainly caused by Gram-negative bacteria, fucoxanthin may impact the reduction of inflammation. Gram-negative bacteria contain lipopolysaccharide (LPS), an endotoxin, which is a membrane component. During infection, LPS affects the inflammatory response, including septic shock, fever, and microbial invasion [42]. It was shown that fucoxanthin inhibited induced by LPS production of pro-inflammatory cytokines (IL-1β, IL-6, and TNF-α) by suppressing the NF-κB activation and the MAPK phosphorylation. Moreover, it reduced the levels of inducible nitric oxide synthase (iNOS) and cyclooxygenase 2 (COX-2) proteins [42–44]. Unfortunately, the direct antibacterial mechanism of fucoxanthin action is not known [12].

In the literature, there is a relationship between the antioxidant and antibacterial properties of natural chemical compounds [37]. The possible mechanisms of antibacterial activity of antioxidants include three basic ways: outer membrane permeability, cytoplasm leakage, and inhibition of nucleic acid formation [45]. The stronger effect of fucoxanthin against Gram-positive than Gram-negative bacteria, exhibited in our research and in the works of other authors [7,12], indicates that this biological activity depends on the differences in the cell wall structure and composition of both types of bacteria.

5. Conclusions

Our investigations confirm the antibacterial properties of fucoxanthin. The obtained results suggest that the above-mentioned substance can be a good antibacterial agent against some Gram-positive pathogens, including *Streptococcus agalactiae, Staphylococcus epidermidis, Staphylococcus aureus,* and weaker against Gram-negative ones (e.g., *Escherichia coli, Klebsiella oxytoca, K. pneumoniae*). On the other hand, it seems that fucoxanthin is not active towards strict anaerobic bacteria.

Author Contributions: Conceptualization, funding acquisition, methodology, visualization, T.M.K.; investigations, writing—original draft, review and editing, T.M.K. and A.A.

Funding: This research received no external funding. This research was paid for by the budget of the Department of Medical Microbiology, Poznań University of Medical Sciences, Poland.

Conflicts of Interest: The authors declare no conflict of interest.

References

1. Mohamed, S.; Hashim, S.N.; Rahman, H.A. Seaweeds: A sustainable functional food for complementary and alternative therapy. *Trends Food Sci. Technol.* **2012**, *23*, 83–96. [CrossRef]
2. Takaichi, S. Carotenoids in algae: Distributions, biosyntheses and functions. *Mar. Drugs* **2011**, *9*, 1101–1118. [CrossRef] [PubMed]
3. Peng, J.; Yuan, J.P.; Wu, C.F.; Wang, J.H. Fucoxanthin, a marine carotenoid present in brown seaweeds and diatoms: Metabolism and bioactivities relevant to human health. *Mar. Drugs* **2011**, *9*, 1806–1828. [CrossRef] [PubMed]
4. D'Orazio, N.; Gemello, E.; Gammone, M.A.; de Girolamo, M.; Ficoneri, C.; Riccioni, G. Fucoxantin: A treasure from the sea. *Mar. Drugs* **2012**, *10*, 604–616. [CrossRef] [PubMed]
5. Gammone, M.A.; Riccioni, G.; D'Orazio, N. Marine carotenoids against oxidative stress: Effects on human health. *Mar. Drugs* **2015**, *13*, 6226–6246. [CrossRef] [PubMed]
6. Mori, K.; Ooi, T.; Hiraoka, M.; Oka, N.; Hamada, H.; Tamura, M.; Kusumi, T. Fucoxanthin and its metabolites in edible brown algae cultivated in deep seawater. *Mar. Drugs* **2004**, *2*, 63–72. [CrossRef]
7. Deyab, M.A.; Abou-Dobara, M.I. Antibacterial activity of some marine algal extracts against most nosocomial bacterial infections. *Egypt. J. Exp. Biol. Bot.* **2013**, *9*, 281–286.
8. Rajauria, G.; Abu-Ghannam, N. Isolation and partial characterization of bioactive fucoxanthin from *Himanthalia elongata* brown seaweed: A TLC-based approach. *Int. J. Anal. Chem.* **2013**, *2013*, 802573. [CrossRef]
9. Jung, H.A.; Ali, M.Y.; Choi, R.J.; Jeong, H.O.; Chung, H.Y.; Choi, J.S. Kinetics and molecular docking studies of fucosterol and fucoxanthin, BACE1 inhibitors from brown algae *Undaria pinnatifida* and *Ecklonia stolonifera*. *Food Chem. Toxicol.* **2016**, *89*, 104–111. [CrossRef]
10. Maeda, H.; Fukuda, S.; Izumi, H.; Saga, N. Anti-oxidant and fucoxanthin contents of brown alga Ishimozuku (*Sphaerotrichia divaricata*) from the West Coast of Aomori, Japan. *Mar. Drugs* **2018**, *16*, 255. [CrossRef]
11. Koo, S.Y.; Hwang, J.H.; Yang, S.H.; Um, J.I.; Hong, K.W.; Kang, K.; Pan, C.H.; Hwang, K.T.; Kim, S.M. Anti-obesity effect of standardized extract of microalga *Phaeodactylum tricornutum* containing fucoxanthin. *Mar. Drugs* **2019**, *17*, 311. [CrossRef] [PubMed]
12. Liu, Z.; Sun, X.; Sun, X.; Wang, S.; Xu, Y. Fucoxanthin isolated from *Undaria pinnatifida* can interact with *Escherichia coli* and lactobacilli in the intestine and inhibit the growth of pathogenic bacteria. *J. Ocean Univ. China.* **2019**, *18*, 926–932. [CrossRef]
13. Silva, A.F.R.; Abreu, H.; Silva, A.M.S.; Cardoso, S.M. Effect of oven-drying on the recovery of valuable compounds from *Ulva rigida, Gracilaria* sp. and *Fucus vesiculosus*. *Mar. Drugs* **2019**, *17*, 90. [CrossRef] [PubMed]
14. Walsh, P.J.; McGrath, S.; McKelvey, S.; Ford, L.; Sheldrake, G.; Clarke, S.A. The osteogenic potential of brown seaweed extracts. *Mar. Drugs* **2019**, *17*, 141. [CrossRef] [PubMed]
15. Fucoxanthin. 2019. Available online: http://www.chemspider.com/Chemical-Structure.21864745.html (accessed on 3 June 2019).
16. Heo, S.J.; Ko, S.C.; K, S.M.; Kang, H.S.; Kim, J.P.; Kim, S.H.; Lee, K.W.; Cho, M.G.; Jeon, Y.J. Cytoprotective effect of fucoxanthin isolated from brown algae *Sargassum siliquastrum* against H_2O_2-induced cell damage. *Eur. Food Res. Technol.* **2008**, *228*, 145–151. [CrossRef]
17. Heo, S.J.; Jeon, Y.J. Protective effect of fucoxanthin isolated from *Sargassum siliquastrum* on UV-B induced cell damage. *J. Photochem. Photobiol. B Biol.* **2009**, *95*, 101–107. [CrossRef] [PubMed]
18. Galasso, C.; Corinaldesi, C.; Sansone, C. Carotenoids from marine organisms: Biological functions and industrial applications. *Antioxidants* **2017**, *6*, 96. [CrossRef]

19. Chen, S.J.; Lee, C.J.; Lin, T.B.; Peng, H.Y.; Liu, H.J.; Chen, Y.S.; Tseng, K.W. Protective effects of fucoxanthin on ultraviolet b-induced corneal denervation and inflammatory pain in a rat model. *Mar. Drugs* **2019**, *17*, 152. [CrossRef]

20. Muradian, K.; Vaiserman, A.; Min, K.J.; Fraifeld, V.E. Fucoxanthin and lipid metabolism: A minireview. *Nutr. Metab. Card. Dis.* **2015**, *25*, 891–897. [CrossRef]

21. Gammone, M.A.; D'Orazio, N. Anti-obesity activity of the marine carotenoid fucoxanthin. *Mar. Drugs* **2015**, *13*, 2196–2214. [CrossRef]

22. Miyashita, K. Function of marine carotenoids. *Forum Nutr.* **2009**, *61*, 136–146. [PubMed]

23. D'Orazio, N.; Gammone, M.A.; Gemello, E.; De Girolamo, M.; Cusenza, S.; Riccioni, G. Marine bioactives. Pharmacological properties and potential applications against inflammatory diseases. *Mar. Drugs* **2012**, *10*, 812–833. [CrossRef] [PubMed]

24. Hosokawa, M.; Wanezaki, S.; Miyauchi, K.; Kurihara, H.; Kohno, H.; Kawabata, J.; Takahashi, K. Apoptosis-inducing effect of fucoxanthin on human leukemia cell HL-60. *Food Sci. Technol. Res.* **1999**, *5*, 243–246. [CrossRef]

25. Kim, K.N.; Heo, S.J.; Kang, S.M.; Ahn, G.; Jeon, Y.J. Fucoxanthin induces apoptosis in human leukemia HL-60 cells through a ROS-mediated Bcl-xL pathway. *Toxicol. In Vitro* **2010**, *24*, 1648–1654. [CrossRef] [PubMed]

26. Hosokawa, M.; Kudo, M.; Maeda, H.; Kohno, H.; Tanaka, T.; Miyashita, K. Fucoxanthin induces apoptosis and enhances the antiproliferative effect of the PPARgamma ligand, troglitazone, on colon cancer cells. *Biochim. Biophys. Acta.* **2004**, *1675*, 113–119. [CrossRef] [PubMed]

27. Kotake-Nara, E.; Kushiro, M.; Zhang, H.; Sugawara, T.; Miyashita, K.; Nagao, A. Carotenoids affect proliferation of human prostate cancer cells. *J. Nutr.* **2001**, *131*, 3303–3306. [CrossRef] [PubMed]

28. Kotake-Nara, E.; Asai, A.; Nagao, A. Neoxanthin and fucoxanthin induce apoptosis in PC-3 human prostate cancer cells. *Cancer Lett.* **2005**, *220*, 75–84. [CrossRef] [PubMed]

29. Satomi, Y. Fucoxanthin induces GADD45A expression and G1 arrest with SAPK/JNK ctivation in LNCap human prostate cancer cells. *Anticancer Res.* **2012**, *32*, 807–813.

30. Zhang, Z.; Zhang, P.; Hamada, M.; Takahashi, S.; Xing, G.; Liu, J.; Sugiura, N. Potential chemoprevention effect of dietary fucoxanthin on urinary bladder cancer EJ-1 cell line. *Oncol. Rep.* **2008**, *20*, 1099–1103. [CrossRef]

31. Rokkaku, T.; Kimura, R.; Ishikawa, C.; Yasumoto, T.; Senba, M.; Kanaya, F.; Mori, N. Anticancer effects of marine carotenoids, fucoxanthin and its deacetylated product, fucoxanthinol, on osteosarcoma. *Int. J. Oncol.* **2013**, *43*, 1176–1186. [CrossRef]

32. Wang, J.; Ma, Y.; Yang, J.; Jin, L.; Gao, Z.; Xue, L.; Hou, L.; Sui, L.; Liu, J.; Zou, X. Fucoxanthin inhibits tumour-related lymphangiogenesis and growth of breast cancer. *J. Cell Mol. Med.* **2019**, *23*, 2219–2229. [CrossRef] [PubMed]

33. Mei, C.; Zhou, S.; Zhu, L.; Ming, J.; Zeng, F.; Xu, R. Antitumor effects of *Laminaria* extract fucoxanthin on lung cancer. *Mar. Drugs* **2017**, *15*, 39. [CrossRef] [PubMed]

34. Yu, R.X.; Hu, X.M.; Xu, S.Q.; Jiang, Z.J.; Yang, W. Effects of fucoxanthin on proliferation and apoptosis in human gastric adenocarcinoma MGC-803 cells via JAK/STAT signal pathway. *Eur. J. Pharmacol.* **2011**, *657*, 10–19. [CrossRef] [PubMed]

35. Ganesan, P.; Matsubara, K.; Sugawara, T.; Hirata, T. Marine algal carotenoids inhibit angiogenesis by down-regulating FGF-2-mediated intracellular signals in vascular endothelial cells. *Mol. Cell Biochem.* **2013**, *380*, 1–9. [CrossRef]

36. Garg, S.; Afzal, S.; Elwakeel, A.; Sharma, D.; Radhakrishnan, N.; Dhanjal, J.K.; Sundar, D.; Kaul, S.C.; Wadhwa, R. Marine carotenoid fucoxanthin possesses anti-metastasis activity: Molecular evidence. *Mar. Drugs* **2019**, *17*, 338. [CrossRef]

37. Shannon, E.; Abu-Ghannam, N. Antibacterial derivatives of marine algae: An overview of pharmacological mechanisms and applications. *Mar. Drugs* **2016**, *14*, 81. [CrossRef]

38. Pérez, M.J.; Falqué, E.; Domínguez, H. Antimicrobial action of compounds from marine seaweed. *Mar. Drugs* **2016**, *14*, 52. [CrossRef]

39. CLSI. *Performance Standards for Antimicrobial Disk Susceptibility Tests. Approved Standard*, 11th ed.; Clinical and Laboratory Standards Institute: Wayne, NJ, USA, 2012; Volume 32.

40. Karpiński, T.M.; Adamczak, A. Antibacterial activity of ethanolic extracts of some moss species. *Herba Pol.* **2017**, *63*, 11–17. [CrossRef]

41. Karpiński, T.M. Efficacy of octenidine against *Pseudomonas aeruginosa* strains. *Eur. J. Biol. Res.* **2019**, *9*, 135–140.

42. Kim, K.N.; Heo, S.J.; Yoon, W.J.; Kang, S.M.; Ahn, G.; Yi, T.H.; Jeon, Y.J. Fucoxanthin inhibits the inflammatory response by suppressing the activation of NF-κB and MAPKs in lipopolysaccharide-induced RAW 264.7 macrophages. *Eur. J. Pharmacol.* **2010**, *649*, 369–375. [CrossRef]

43. Heo, S.J.; Yoon, W.J.; Kim, K.N.; Oh, C.; Choi, Y.U.; Yoon, K.T.; Kang, D.H.; Qian, Z.J.; Choi, I.W.; Jung, W.K. Anti-inflammatory effect of fucoxanthin derivatives isolated from *Sargassum siliquastrum* in lipopolysaccharide-stimulated RAW 264.7 macrophage. *Food Chem. Toxicol.* **2012**, *50*, 3336–3342. [CrossRef] [PubMed]

44. Jiang, X.; Wang, G.; Lin, Q.; Tang, Z.; Yan, Q.; Yu, X. Fucoxanthin prevents lipopolysaccharide-induced depressive-like behavior in mice via AMPK-NF-κB pathway. *Metab. Brain Dis.* **2019**, *34*, 431–442. [CrossRef] [PubMed]

45. Naqvi, S.A.R.; Nadeem, S.; Komal, S.; Naqvi, S.A.A.; Mubarik, M.S.; Qureshi, S.Y.; Ahmad, S.; Abbas, A.; Zahid, M.; Khan, N.U.H.; et al. *Antioxidants: Natural Antibiotics*, 1st ed.; IntechOpen: London, UK, 2019; pp. 1–17.

Review

antioxidants

MDPI

Do We Utilize Our Knowledge of the Skin Protective Effects of Carotenoids Enough?

Anamaria Balić [1],* and Mislav Mokos [2]

[1] Department of Dermatology and Venereology, University Hospital Centre Zagreb, School of Medicine, University of Zagreb, Šalata 4, 10 000 Zagreb, Croatia
[2] School of Medicine, University of Zagreb, Šalata 3, 10 000 Zagreb, Croatia
* Correspondence: jovicanamaria@gmail.com

Received: 30 June 2019; Accepted: 24 July 2019; Published: 31 July 2019

Abstract: Due to their potential health-promoting effects, carotenoids have drawn both scientific and public attention in recent years. The primary source of carotenoids in the human skin is diet, mainly fruits, vegetables, and marine product, but they may originate from supplementation and topical application, too. In the skin, they accumulate mostly in the epidermis and act as a protective barrier to various environmental influences. Namely, the skin is exposed to numerous environmental factors, including ultraviolet radiation (UVR), air pollution, and smoking, that cause oxidative stress within the skin with consequent premature (extrinsic) aging. UVR, as the most prominent environmental factor, may cause additional detrimental skin effects, such as sunburn, DNA damage, and skin cancer. Therefore, photoprotection is the first line intervention in the prevention of premature aging and skin cancer. Numerous studies have demonstrated that carotenoids, particularly β-carotene, lycopene, lutein, and astaxanthin, have photoprotective effects, not only through direct light-absorbing properties, but also through their antioxidant effects (scavenging reactive oxygen species), as well as by regulation of UV light-induced gene expression, modulation of stress-dependent signaling, and/or suppression of cellular and tissue responses like inflammation. Interventional studies in humans with carotenoid-rich diet have shown its photoprotective effects on the skin (mostly by decreasing the sensitivity to UVR-induced erythema) and its beneficial effects in prevention and improvement of skin aging (improved skin elasticity and hydration, skin texture, wrinkles, and age spots). Furthermore, carotenoids may be helpful in the prevention and treatment of some photodermatoses, including erythropoietic protoporphyria (EPP), porphyria cutanea tarda (PCT) and polymorphous light eruption (PMLE). Although UVR is recognized as the main etiopathogenetic factor in the development of non-melanoma skin cancer (NMSC) and melanoma, and the photoprotective effects of carotenoids are certain, available studies still could not undoubtedly confirm the protective role of carotenoids in skin photocarcinogenesis.

Keywords: antioxidant; skin health; skin aging; skin cancer; photocarcinogenesis; oral photoprotection; nutraceuticals; cosmeceuticals

---

1. Introduction

Skin acts as a protective barrier against various environmental influences such as mechanical damage, noxious substances, microorganisms, free radicals, and ultraviolet radiation (UVR). In addition to endogenous factors, external stressors, primarily UVR, result in the alterations of the skin such as inflammation, impaired immune function and epidermal barrier homeostasis, photoaging, and most importantly the formation of various skin diseases and malignancies [1,2]. Due to the diverse proven health-promoting effects, carotenoids are along with some other nutraceuticals widely investigated and put in the focus of interest of many scientific and health-promoting groups [3–7]. We have also noticed a daily increase in the number of analytical techniques for carotenoids determination [8,9]. Due to their vast abundance, especially in plant-derived food, they are an integral part of the human diet. Although carotenoids have never been identified as being essential for humans as long as preformed vitamin A is available through the diet [10,11], the increasing evidence of their important role in the biology and human health continues to stimulate broad interest in the carotenoid field. Consumption of food products and supplements containing carotenoids has increased tremendously due to their health- and skin-favorable or disease-preventive effects, especially in UVR protection and consequently in the prevention of photo-induced dermatoses and skin aging (Figure 1).

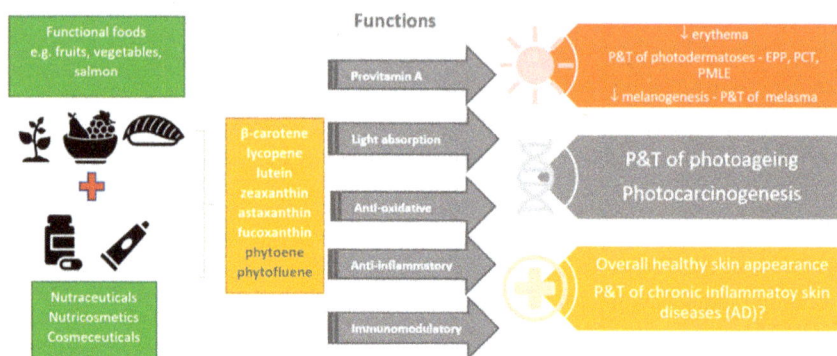

Figure 1. Diverse skin health-promoting effects of food rich in carotenoids or their supplementation through nutraceuticals or cosmeceuticals. P&T, prevention and treatment; EPP, erythropoietic protoporphyria; PCT, porphyria cutanea tarda; PMLE, polymorphous light eruption; AD, atopic dermatitis.

2. Carotenoids

Carotenoids are known as fat-soluble plant pigments widely distributed in nature that provide diverse colors such as yellow, red, and orange to fruits and vegetables [12]. While they are biosynthesized primarily by plants and algae, as well as by fungi and bacteria, we can find carotenoids throughout the animal kingdom and humans due to selective absorption along the food chain [13]. These lipophilic molecules are based on the chemical structure classified as carotenes and xanthophylls [14]. Both classes have a common C40 polyisoprenoid structure containing a series of centrally located, conjugated double bonds which act as a light-absorbing chromophore. Carotenoids that exist as pure non-polar hydrocarbons are referred to as carotenes (α-carotene, β-carotene, and lycopene); on the contrary, xanthophylls (β-cryptoxanthin, lutein, zeaxanthin, astaxanthin) are more polar carotenoids that contain oxygen as a functional group in its structure either as a hydroxyl or keto group as the end group [15]. The presence of a polar group in the structure affects the polarity and biological function of the compounds [16]. The main sources of around 50 carotenoids in the human diet are fruits and vegetables, followed by green leaves and, to a minor extent, some marine products [14,16–18]. The transport of carotenoids from the gut occurs on the uptake with chylomicrons into the lymph, followed by the circulation of lipoprotein particles in the blood, and then transportation to various target tissues with large interorgan differences [19,20]. Absorption of carotenoids in the gut is shown to be mediated by both simple diffusion, which is dependent on the concentration gradient, and facilitated diffusion through cholesterol membrane transporters such as scavenger receptor class B member 1 (SR-B1) and a cluster of differentiation 36 (CD 36) [21,22]. Till now, more than 800 carotenoids have been identified, but only several are found in the human organism, including α-carotene, β-carotene, lutein, and lycopene, as well as the zeaxanthin, and α- and β-cryptoxanthin [23–26]. Among 800 known carotenoids only 20 of them are studied in sufficient depth; thus, it is an area with huge amount of space in front of scientists. Both α- and β-carotenes, and β-cryptoxanthin are provitamin A carotenoids with different percentage of provitamin A activity [27]. Among provitamins, β-carotene is the most common carotenoid in the diet of mammals and has the highest conversion efficiency to vitamin A with no difference between naturally occurring and chemically synthesized β-carotene. Some of the absorbed β-carotene is cleaved by the enzyme β, β–carotene-15,15′-monooxygenase 1 (BCMO1) into two molecules of all-*trans*-retinal which can be either further reduced reversibly or oxidized. Various fruits and vegetables are rich in carotenoids [28–34], especially lycopene, such as tomatoes, asparagus, pink grapefruit, guava, and watermelon. Pumpkin, carrots, sweet potatoes, mangos, and papaya are some examples of β-carotene containing food. Oranges, tangerines, nectarines, mango, and papaya are rich in cryptoxanthin. We can find lutein and zeaxanthin in leafy green vegetables, pumpkins, and red peppers. Marine food like microalgae, yeast, salmon, trout, krill, shrimp, crayfish, and crustacea are known as sources of astaxanthin, a strong photoprotectant which has been attributed an enormous potential for protecting the organism against a wide range of diseases due to its strong antioxidant and anti-inflammatory effects [35–40]. We also need to mention fucoxanthin, another marine carotenoid with remarkable biological properties that is found in the marine macroalgae—brown seaweeds, and the microalgae—diatoms [41].

The way of food processing affects available carotenoid contents [42,43]. Their bioavailability varies from ~10% in raw materials up to about 50% in commercial and oil-based products [16]. For example, the content of available lycopene is higher in cooked tomatoes and is also increased by the addition of oil, such as olive oil [12,44]. The biological properties of carotenoids are manifold. Besides their natural role as pigments, provitamins, and photosynthetic organisms, they have been demonstrated to possess numerous health-promoting effects as being efficient antioxidants (AOs) – by decreasing reactive oxygen species (ROS) [3,6,45]. However, some carotenoids, especially highly concentrated carotenes, showed pronounced pro-oxidative effects [17,46,47]. It is known that singlet oxygen quenching ability of various carotenoids in organic solvents increases with an increasing number of conjugated double bonds [48]. The biologically important shorter chain C40 carotenoids have quenching rate constants half those of longer chain carotenoids such as β-carotene; e.g., in the case of lutein with ten double bonds the quenching rate is lower than in those with 11 double bonds such as β-carotene or zeaxanthin [49].

3. Carotenoids and the Skin

The amount of carotenoids in the skin depends on dietary intake or supplementation, and their bioavailability from various food [50]. After absorption in the gut and transportation into the skin, carotenoids accumulate mainly into the epidermis. It is thought that they are transported into the epidermis by the same prior mentioned cholesterol transporters—SR-B1 found in the basal layers because the epidermis is an active site of cholesterol accumulation crucial for the epidermal barrier function [51]. Due to the accumulation of carotenoids in the epidermis, in the cases of excessive carotenoid intake, mostly β-carotene, carotenoderma occurs as yellowish discoloration of the skin being most obvious on the palms and soles [52,53]. Besides the oral route of carotenoid administration, topical application is just as important, being especially of great interest for cosmetic companies [54–58]. Topical application of antioxidant (AO) substances such as carotenoids is closely related to skin protection from environmental factors and anti-aging [59,60]. To determine the total dermal carotenoid level in comparison with their respective plasma levels, Scarmo et al. [61] performed skin biopsies of healthy individuals and collected their blood samples for correlation of individual and total carotenoid content using high-performance liquid chromatography (HPLC). Later on, this group of authors in their studies used less invasive methods such as a resonance Raman spectroscopy (RRS) to assess carotenoid status in human tissues [62]. Based on their findings, it is suggested that β-carotene and lycopene are present in greater abundance in human skin, in comparison to zeaxanthin and lutein, possibly indicating a specific role of carotenes in human skin photoprotection. Interestingly, levels of carotenoids in skin differ within the skin layers and the body locations with the highest levels in the skin of palms, forehead, and dorsal skin [63]. The lifestyles of individuals also reflect carotenoid levels in the skin; precisely, stressors like UVR, illness, smoking, and alcohol consumption lower their concentrations [64]. As more and more topical products and oral supplements with antioxidative effects are occurring on the market, Darvin et al. wanted to examine in vivo whether the topical, systemic or combined application of AOs/carotenoids is effective in increasing their concentration in the skin by using non-invasive RRS [65]. In this study, carotenoids were applied systemically (carotenoid tablets) at physiological concentrations like those contained in a healthy diet, and topically (cream) in the concentrations corresponding to those physiologically present in healthy skin. Results of this study showed a statistically significant increase of AOs levels in human skin with all forms of treatments—topical, systemic, and combined. A combination of topical and systemic AOs induced the highest accumulation in the skin, which suggested that combined treatment might be an optimal form of protection of the human skin. Carotenoid levels after the end of treatment were preserved for about two weeks following the topical application, and up to five weeks after systemic administration. These results are explained by the fact that topically applied AOs are stored in *stratum corneum* for a short time only due to their rapid depletion by skin desquamation, textile contact, washing, and environmental stress. On the other hand, the systemically applied carotenoids are stored in body fat tissue and slowly release onto the skin surface with sweat and sebum. Based on these findings, it could be concluded that the combined topical and systemic application of carotenoids/AOs represents an optimal form of skin protection, but we need to be aware of the importance of choosing appropriate non-lipid formulation of topical product which does not saturate reservoir of *stratum corneum*, thus allowing the systemically applied carotenoids to penetrate back into the skin.

4. Carotenoids in Skin Photoprotection

People are constantly exposed to ultraviolet (UV) light, some less, some more, depending on the place of living, activities, what they do for a living, hobbies, culture, but also their knowledge of the importance of sun protection and its implementation. It has been estimated that the exposure to solar UV light is ~10% of the total available annual UVR for outdoor workers and ~3% for indoor-working adults [66]. Besides its few beneficial health effects, including vitamin D3 synthesis, improvement of mood through production of endorphins, efficacy in the treatment of various skin diseases, such as psoriasis, vitiligo, and atopic dermatitis; UVR causes many detrimental skin effects—sunburn, ocular damage, photoaging, immune suppression, DNA damage and skin cancer [66–70]. Sun exposure results in photoaging—solar elastosis, skin roughness, furrows, and wrinkles by the mechanisms of mitochondrial deletion and remodeling of the extracellular matrix (ECM) mediated through matrix metalloproteinases (MMP) which cause the damage of collagen and elastin fibers [71]. Most importantly, UVR plays the main role in photo-induced carcinogenesis, melanoma, and non-melanoma skin cancer occurrence [69]. Most of the harmful effects of UVR are mainly mediated by oxidative stress which alters signal transduction pathways such as the nuclear factor-kappa beta (NF-κB)/p65mitogen-activated protein kinase (MAPK), the janus kinase (JAK), signal transduction and activation of transcription (STAT), and the nuclear factor erythroid 2-related factor 2 (Nrf2), causing the damage to biomolecules and affecting the integrity of skin cells leading to skin damage [71]. UVR also induces pro-inflammatory genes and causes immunosuppression by reducing the number and activity of the epidermal Langerhans cells [72].

Many skin diseases form as a result of pathological processes induced by photo-oxidative damage. UVA radiation, which contributes to up to 95% of total UV radiation, does not interact with DNA; however, it is considered the most important source of oxidative stress in human skin. As UVA radiation penetrates the deeper dermis, it plays a significant role in photoaging [73]. On the other hand, UVB radiation is mainly absorbed by keratinocytes in the epidermis and interacts directly with DNA, causing mutations and skin cancer [69,71]. UVB is the leading cause of sunburn, erythema resulting from an inflammatory response to the photodamage of the skin [74]. In the last decade it has been discovered that visible light (400–700 nm) causes solar erythema, thermal damage, induces the melanogenesis in human skin [75] but also contributes to signs of premature photoaging by inducing production of ROS, proinflammatory cytokines, and MMP-1 expression [76]. Additionally, visible light exposure is related to the pathogenesis of some photodermatoses [77].

Photoprotection, either mechanical or pharmacological, is the first line in the prevention of photoaging and skin cancer. Pharmacological photoprotection can be topical or systemic. The main principle of photoprotection is the direct absorption of UV light using suitable compounds. There is an increasing interest in the area of skin protection from UV and visible light by additional endogenous protection by dietary micronutrients with AO properties such as carotenoids, vitamin C and E, and polyphenols [78–81]. Skin photoprotection by nutritional means [54,63,82–85] or topically-applied phytochemicals has been examined by various authors [86,87]. Human interventional studies have documented photoprotective effects of many carotenoids, particularly β-carotene, but also lycopene, lutein, and astaxanthin, provided through topical application, or orally, either by a carotenoid-rich diet or by supplementation, but rather long treatment periods with a minimum of 10 weeks were required [37,47,59,79,88,89]. Most of the carotenoids exhibit absorbance maximum at wavelengths in the range of visible light. However, noncolored carotenoids phytoene and phytofluene have high UV absorption maxima that cover both UVB and UVA range. Lutein and zeaxanthin may protect the skin from blue light, which makes them useful in the prevention or treatment of melasma and in ocular protection. [90–93]. Besides direct light-absorbing properties, carotenoids and some other micronutrients provide endogenous photoprotection and contribute to the prevention of UV damage in humans mostly by their well-known AO effects—scavenging ROS, including excited singlet oxygen and triplet state molecules which would lead to photoinactivation of AO enzymes, lipid peroxidation, and DNA damage induction [63,71,78,94]. Additionally, they interfere with UV light-induced gene expression by multiple pathways, modulate stress-dependent signaling, and/or suppress cellular and tissue responses like inflammation [63,95–97]. The idea of endogenous photoprotection implies that the active compound is available in sufficient amounts at the target site [63]. That is why structural features of carotenoids are important because they influence pharmacokinetic parameters like absorption, distribution, and metabolism and affect the level of the active compound in the skin [11,98,99].

Interventional studies in humans with carotenoid-rich diet have shown its photoprotective effects on the skin, mostly by decreasing the sensitivity to UV radiation-induced erythema. See Table 1.

Table 1. Carotenoids in skin health, repair, and disease—summary.

Carotenoid	Food Source	Photoprotective Effects	Role in Photo-Induced Carcinogenesis Prevention	Role in Photoaging Prevention	Additional Benefits
β-carotene	Pumpkin, carrots, sweet potatoes, mangos, papaya, bilberry [28–30,33]	Prevention of UV-induced erythema [88,100,101], ↑[1] MED[2] [102,103], ↓[3] of the rate of mitochondrial mutation in human dermal fibroblasts after UV irradiation [96]	Delayed tumor appearance and reduced tumor growth rates [104], inhibition of photocarcinogenic enhancement by benzopyrene [105], in vitro induction of apoptosis of melanoma cells by activation of caspase-3,-8, and -9 [106] or by additional regulation of Bcl-2, p53 [107]. In vivo no influence, positive or negative, on the incidence of malignant skin neoplasms, including melanoma. and NMSC[4] [108–110]	O2 quenching ↓MMP[5]-1, -3, and MMP-10 [96,111], ↓MMP-9 partly, by inhibiting Chol-OOHs[6] formation [112]	Combination of β-carotene, lycopene and *Lactobacillus johnsonii* inhibits PMLE[7] [113], protective role in the treatment of EPP[8] and PCT[9] by membrane protection against protoporphyrin IX and uroporphyrin I [114–116]
Lycopene	Tomatoes, asparagus, pink grapefruit, guava, watermelon, peaches, papaya [28,29,33]	Protection against UV-induced erythema [95,117–119], ↓ HO-1[10] ↓ ICAM-1[11] [79]	Inhibits mtDNA deletion [87], inhibits skin tumor formation [120], induction of apoptosis [121], chemoprevention properties in photocarcinogenesis remain contradictory [122]	↓ MMP [79]. ↓MMP-1 and ↓ reduction in fibrillin-1 [87], ↓ amount of furrows and wrinkles [101,123]	PMLE prevention [113], protective role in EPP [115]
Astaxanthin	Microalgae, yeast, salmon, trout, krill, shrimp, crayfish and crustacea [28,29,35,39]	Protection against UV-induced erythema, ↑MED, activation of Nrt213/HO-1 AO pathway [35,37,38,124]	Inhibition of skin cancer and tyosinase in rat model [125]; apoptosis [126]; AO effect, effect on gap junctional communication important for homeostasis, growth control, and development of cells [127–129], may enhance immune responses and potentially exert antitumor activity [130]	↓ wrinkle parameters [131,132], ↑ elasticity, improved skin texture, and ↓ TEWL12 [38,131,133,134], ↓ size of age spots [131], ↑ procollagen type I, ↓MMP-1, -3, -12, also MMP-13 [26,133,135], ↓ malondialdehyde; ↓ residual skin surface components [136,137], ↓ IL14-1α [132], ↓ MIF15, IL-1β, TNF-α16, preserves trans-UCA17 levels [126], ↓ mast cells [135]	Anti-inflammatory properties - ↓ iNOS18, COX-2, and inhibition of NFκB signaling [138]; ↓ TNF-α, IL-1β, IL-6— possible implication for the treatment of inflammatory diseases such as atopic dermatitis [138] and psoriasis. Accelerates wound healing—↓iNOS, ↑wound healing biological markers including Col1A121 and bFGF22 [139]
Lutein/ Zeaxanthin	Leafy green vegetables, peas, broccoli, pumpkins, corn, red peppers, egg yolk, bilberry [28–30,34]	↓ skin edema and erythema after UVR [140], ↓ masts cells number [141], ↓ melanogenesis [93], blocking of eye damage induced by blue light [90,91]	↑ tumor-free survival time, ↓ tumor volume and multiplicity [141], ↓ PCNA[23] and BrdU + epidermal cells [140], reduced incidence of SCC[24] in persons who had a history of skin cancer at baseline [142]	↓overexpression of HO-1, ICAM-1, MMP-1 Genes [79], ↓ MMP-1 and MMP-7, ↑ TIMP-2 [141,143], ↑ surface lipids, skin hydration, and skin elasticity [54,144]	Prevention of melasma, skin-lightening effects [93]

[1] increase, [2] minimal erythema dose, [3] decrease, [4] non-melanoma skin cancer, [5] matrix metalloproteinase, [6] cholesterol hydroperoxides, [7] polymorphous light eruption, [8] erythropoietic protoporphyria, [9] porphyria cutanea tarda, [10] heme oxygenase-1, [11] intercellular adhesion molecule, [12] transepidermal water loss, [13] nuclear factor erythroid 2-related factor 2, [14] interleukin, [15] macrophage migration inhibitory factor, [16] tumour necrosis factor-alpha, [17] trans-urocanic acid, [18] inducible nitric oxide, [19] cyclooxygenase-2, [20] nuclear factor-kappa beta, [21] collagen type I alpha 1 chain, [22] fibroblast growth factor, [23] proliferating cell nuclear antigen, [24] squamous cell carcinoma, [25] tissue inhibitor of metalloproteinase-2.

4.1. Lycopene

Lycopene is considered the most efficient dietary carotenoid when it comes to quenching singlet oxygen in organic solvents [145]. Its quenching efficacy is, in this case, greater than all C40 carotenoids and twice greater than the one of β-carotene. On the contrary, others state that lycopene is only slightly more efficient than β-carotene and that in more biomimetic environments, such as micelles and liposomes, lycopene, and β-carotene have rather similar quenching abilities [49]. There are also human cell protection studies of lycopene on oxidant-induced damage, demonstrating its beneficial AO effects and strong protection role in comparison with other carotenoids [146–148].

Stahl et al. [149] conducted an interventional study to investigate whether intervention with a natural dietary source rich in lycopene protects against UV-induced erythema in humans. They found that ingestion of tomato paste (40 g per day, equivalent to 16 mg lycopene per day) with 10 g of olive oil over ten weeks led to 40% reduction of skin erythema induced by exposure to solar-simulating UVR. No significant protection was found after four weeks of dietary intervention, but after ten weeks, erythema was significantly lower than in the control group receiving olive oil only. Rizwan et al. [87] previously conducted a similar study where they also examined whether ingestion of 55 mg tomato paste in olive oil daily over 12 weeks can protect human skin against UVR-induced effects—erythema, changes in ECM, and mitochondrial DNA (mtDNA) damage. UVR-induced erythema was assessed visually as the minimal erythemal dose (MED) but also quantitatively with an instrument pre-and-post nutrition rich in lycopene. To demonstrate UVR-induced ECM changes, and mtDNA damage, partially mediated by oxidative stress, they performed skin biopsies from unexposed and UVR-exposed skin before and after the nutritional intervention. Skin samples were further analyzed immunohistochemically for MMP-1, fibrillin-1, and procollagen I, and by quantitative polymerase chain reaction (PCR) for mtDNA bp deletion. Based on these and previously-mentioned study results, it is reasonable to conclude that the consumption of food rich in lycopene protects against acute and potentially longer-term aspects of photodamage.

4.2. Lutein

Lutein is also an efficient singlet oxygen quencher, though it is less efficient than lycopene and β-carotene [145,150]. Besides the well-examined photoprotective role of lycopene, Grether-Beck et al. [79] wanted to investigate the skin protective effects of lutein against UVR on a molecular basis. Their double-blind, randomized, controlled study added a fact that besides lycopene-rich tomato nutrient complex (TNC), lutein protects from UVA/B- and UVA1-induced gene expression in human skin. Assuming the role of heme-oxygenase 1, intercellular adhesion molecule 1 (ICAM-1) and MMP-1 mRNA as indicators of oxidative stress, photodermatoses, and photoaging, these study results indicate that TNC and lutein could protect against UVR-induced skin damage.

4.3. β-Carotene

Most studies that describe the role of carotenoids in photoprotection investigated the photoprotective role of β-carotene and its effectiveness in the prevention of UV-induced erythema formation, being especially useful in the treatment and prevention of some photodermatoses, namely EPP and PMLE [83,88,89,100,113,114,151,152]. Systemic photoprotective effects of this provitamin depend both on the dose and the duration of treatment. In most of the interventional studies with carotenoids, photoprotection was observed only after a minimum of 10 weeks of dietary intake or supplementation, with doses >12 mg/day [89,101,153]. A sufficiently long period of treatment is needed to provide optimal photoprotection of the skin. The photoprotective role of β-carotene is fortified also in vitro by the findings that its supplementation significantly reduces the rate of mtDNA mutation in human dermal fibroblasts after UVR [117]. However, the need for caution with isolated β-carotene supplementation was pointed up after human interventional trials that had demonstrated potentially harmful effects of high dosages of this carotenoid and raised a discussion on suitable dose amounts for photoprotection. In two long-term interventional trials in individuals at high risk for cancer (cigarette smokers and asbestos workers) who received β-carotene for several years at doses of 20 and 30 mg/day, there was an ~20% increase in the incidence of lung cancer [6,154]. The authors concluded that the effects of higher doses of β-carotene in cancer pathogenesis could partly be explained by the formation of eccentric cleavage products of β-carotene, which can interfere with the retinoic acid receptor-mediated signaling pathway [27]. Despite these interventional trials results, dermatologists are still encouraged to recommend β-carotene supplements for photoprotection, especially in patients with EPP, photosensitive diseases, and to reduce the phototoxic damage caused by some drugs [155].

4.4. Astaxanthin

In recent years, much attention has been put on the health and skin benefits of astaxanthin [35]. Astaxanthin, a marine pigment, is mostly produced by the microalga *Haematococcus pluvialis* to protect its cells from sun radiation, UV-light, and oxidation [156]. Camera et al. [37] conducted a study in which they examined the modulation of UVA-related injury by astaxanthin, canthaxanthin, and β-carotene for systemic photoprotection in human dermal fibroblast. In this study, astaxanthin showed significant photoprotective effect and counteracted UVA-induced alterations to a greater extent. The uptake of astaxanthin by dermal fibroblasts was higher than that of other two carotenoids, which led to the assumption that the antioxidative photoprotective effect of astaxanthin was stronger than of the other substances. Other in vitro studies fortified the photoprotective role of astaxanthin by showing that it could interfere with UVA-induced MMP-1 and skin fibroblast elastase/neutral endopeptidase expression [135,157].

4.5. Fucoxanthin

Similar to astaxanthin, orange-colored pigment accumulated by marine plants, which has AO and provitamin A effects, fucoxanthin, shows a protective effect against UVB-induced skin damage by decreasing intracellular ROS [40,158]. Matsui et al. suggest that this skin sun-protective effect may be due to the restoration of filaggrin and promotion of skin barrier formation, unrelated to AO action [159]. As inherited or acquired filaggrin deficiency substantially contributes to the pathogenesis of atopic dermatitis, fucoxanthin might be useful in its and similar conditions treatment. Besides the sun-protective effect, fucoxanthin exhibits anti-pigmentary activity in UVB-induced melanogenesis either by oral or topical route of application presumably by the suppression of melanogenic stimulant receptors and prostaglandin E2 synthesis [160].

5. Carotenoids and Photocarcinogenesis

Based on the above statements of the skin photoprotective role of carotenoids, a question about their relationship with skin cancer incidence pops up. Regardless of their proved role as agents that prevent skin cancer in vitro and in animal studies; human interventional or epidemiological studies regarding the effect of carotenoids on the incidence of UV-induced skin cancer are lacking (see Table 1). However, the study conducted by Heinen et al. [142] among Australian population showed that high dietary intake of lutein and zeaxanthin was related with a decreased incidence of SCC in persons who had a history of skin cancer.

6. Carotenoids and Skin Aging

A healthy diet based on large amounts of fruits and vegetables is known to be beneficial in the prevention of skin aging, especially photoaging, as it increases the concentration of AOs in the blood and the skin substantially [65,161–164].

6.1. Lycopene

Meinke et al. [164] measured the blood and skin levels of the carotenoids in individuals after oral administration of natural kale extract, or placebo oil, for four weeks. Carotenoid bioaccessibility was evaluated using RRS on the forehead skin and the palm. For the analysis of the blood serum, the standard HPLC method was employed. In this study, carotenoids' bioaccessibility increased significantly in both skin and serum, but increases in the skin were delayed when compared with serum levels and depended on the dermal area as well as on the type of carotenoid. Lycopene bioaccessibility increased more in the skin compared to the blood, which indicates that the natural kale extract stabilizes the AO network in the skin. Carotenoids' levels decreased significantly faster in the blood than in the skin after the end of supplementation, which may indicate a peripheral buffer function of the skin for carotenoids. It is shown that individuals with a higher concentration of lycopene in the skin have a significantly smaller amount of wrinkles and furrows than individuals with lower concentrations which fortifies the protective role of lycopene when it comes to pro-oxidative damage [101,123]. It has also been demonstrated that the skin roughness is reduced after systemic application of carotenoids [44,165].

6.2. Lutein and Zeaxanthin

Some studies have examined the efficacy of lutein and zeaxanthin, found naturally in the skin, upon several skin physiology parameters. A group of Italian authors [54] designed randomized, double-blind, placebo-controlled, 12-week clinical multicenter study to evaluate the effect of lutein and zeaxanthin administered both orally and topically upon human skin of forty healthy middle-aged women that exhibited signs of premature skin aging. The study results showed that the provision of the previous two carotenoids ensures multiple benefits to the skin. Besides the prevention of UVR-inducible damage, these xanthophylls also improved skin features—skin hydration, elasticity, and increased skin surface lipids. These beneficial skin effects were achieved regardless of whether both xanthophylls were administered topically or orally, which demonstrates that the simultaneous administration of these carotenoids by both routes could result in greater skin health. Similar findings were observed in a study observing the skin effects of only zeaxanthin both topically and/or orally [144]. Meinke et al. recently fortified their prior finding of the beneficial role of carotenoids in the prevention of skin aging by showing that a natural carotenoid-rich extract could prevent the aging-related collagen I degradation in the dermis and improve the ECM [166]. In their study, 29 healthy middle-aged female volunteers received a supplement in the form of a carotenoid-rich natural curly kale extract containing 1650 µg of carotenoids in total (three capsules of 550 µg), once a day for up to 10 months. Their skin was examined in vivo using noninvasive RS-based scanners and two-photon tomography for determination of skin carotenoids before, after five months, and after ten months of daily supplementation. The results showed a significant increase in skin carotenoids and the collagen I/elastin aging index of the dermis proportional to the duration of carotenoid supplementation.

6.3. β-Carotene

Compared to a vast number of experimental studies investigating the effectiveness of β-carotene in the prevention of UV-induced erythema and skin photodamage, there are only a few clinical studies investigating its role in photoaging. One of them is a Korean study in which was conducted with the aim to determine the effects of 90 days supplementation with two different doses of dietary β-carotene (30 and 90 mg/day) on UV-induced DNA damage in human skin in vivo, procollagen type I, MMP-1, and fibrillin-1 gene expression, and skin elasticity and wrinkle formation [167]. Their findings led to a conclusion that low-dose β-carotene supplementation prevents and repairs photoaging, which reflects as improvement in facial wrinkles and elasticity. However, other clinical studies have failed to convincingly demonstrate its beneficial effects [168].

6.4. Fucoxanthin

Fucoxanthin may be an active ingredient of cosmeceuticals and nutraceuticals used in the protection of the skin from photoaging [169]. Its beneficial role in skin aging protection is based on the findings of Urikura et al. [170] which demonstrate that fucoxanthin significantly suppressed UVB-induced wrinkle formation, epidermal hypertrophy, MMP-13, vascular endothelial growth factor (VEGF) and the increase of other reactive substances in the UVB-irradiated animal model of hairless mice.

6.5. Astaxanthin

As we mentioned in the previous section, astaxanthin is among the carotenoids considered to be a potent skin protective nutrient due to its natural capacity to protect cells from irradiation and oxidation, proven to prevent or minimize the signs of UVB-induced skin damage but also UVA-induced photoaging such as skin wrinkling or sagging by topical or oral administration routes [35,37,133,157]. Several human studies, which were later confirmed in animal studies [135,171], demonstrated that astaxanthin, in addition to improvement of the appearance of wrinkles, also improved skin elasticity, moisture, age spots, and skin texture [172,173]. In vitro studies have demonstrated that astaxanthin improves the function of mitochondria and has protective effects on human skin fibroblasts by exhibiting other biological functions rather than AO, including effects on gap junctional communication important for homeostasis, growth control, and development of cells [174]. In that way, it can protect skin cells from ROS and preserve the collagen, which results in the smooth and youthful appearance of the skin. Based on the results of numerous studies, it is reasonable to conclude that astaxanthin supplementation has promising functional improvements to the skin and that it can help reduce the skin aging process (see Table 1).

7. Future Perspectives

Available data on the protective effects of carotenoids on human skin may encourage their implementation in the field of dermatology as nutraceuticals, cosmeceuticals, and photoprotectants. Due to their antioxidant, anti-inflammatory, and immunomodulatory effects, the optimal supply of these micronutrients increases dermal defense against UVR, maintains longer-term protection, alleviates certain photodermatoses, and contributes to better skin health and appearance. Hopefully, besides their beneficial role in reducing late effects of UVR such as photoaging, carotenoids may be helpful in the treatment of inflammatory skin diseases like atopic dermatitis and psoriasis. If so, their use could be beneficial at a population level or at least in those perceived to be at high risks, such as outdoor workers, immunosuppressed patients, patients with various photodermatoses other than PMLE, EPP, and porphyria cutanea tarda, but also ones requiring repeated courses of phototherapy. In practice, advice on achieving a proper intake of protective carotenoids should probably emphasize natural food sources rather than supplements. As the diet rich in fruits and vegetables is already being encouraged in many countries due to its various protective health effect, it should be a public health message in each state. As nanotechnology is a promising field of research for the development of nutrient delivery systems, future interventions in the area of carotenoid micronutrition and cosmetics would be of benefit for dermatology patients. Efforts are also needed to enhance the knowledge of already known and still undiscovered health-promoting effects of various carotenoids along with other antioxidants, but also in the development of new topical and systemic photoprotective drugs. We hope that the application and advancements in nutritional and topical skin protection, and in new technologies will enable us to utilize more knowledge of the skin protective effects of carotenoids when approaching our patients.

Author Contributions: A.B. and M.M. contributed actively to the preparation of the manuscript. Conceptualization, A.B.; Writing—original draft preparation, A.B., M.M.; Writing—review and editing, A.B.; Table creation, M.M.; Figure creation, A.B.; Collecting sources of information, A.B., M.M.; Supervision, A.B.

Funding: This research received no external funding.

Conflicts of Interest: The authors declare no conflict of interest.

References

1. Feingold, K.R.; Denda, M. Regulation of permeability barrier homeostasis. *Clin. Dermatol.* **2012**, *30*, 263–268. [CrossRef]

2. Del Rosso, J.Q.; Cash, K. Topical corticosteroid application and the structural and functional integrity of the epidermal barrier. *J. Clin. Aesthet. Dermatol.* **2013**, *6*, 20–27.

3. Milani, A.; Basirnejad, M.; Shahbazi, S.; Bolhassani, A. Carotenoids: Biochemistry, pharmacology and treatment. *Br. J. Pharmacol.* **2017**, *174*, 1290–1324. [CrossRef]

4. Stahl, W.; Sies, H. Bioactivity and protective effects of natural carotenoids. *Biochim. Biophys. Acta Mol. Basis Dis.* **2005**, *1740*, 101–107. [CrossRef]

5. Schieber, A.; Weber, F. Carotenoids. *Handb. Nat. Pigment. Food Beverages* **2016**, 101–123.

6. Krinsky, N.I.; Johnson, E.J. Carotenoid actions and their relation to health and disease. *Mol. Aspects Med.* **2005**, *26*, 459–516. [CrossRef]

7. Woodside, J.V.; McGrath, A.J.; Lyner, N.; McKinley, M.C. Carotenoids and health in older people. *Maturitas* **2015**, *80*, 63–68. [CrossRef]

8. Amorim-Carrilho, K.T.; Cepeda, A.; Fente, C.; Regal, P. Review of methods for analysis of carotenoids. *TrAC Trends Anal. Chem.* **2014**, *56*, 49–73. [CrossRef]

9. Abar, L.; Vieira, A.R.; Aune, D.; Stevens, C.; Vingeliene, S.; Navarro Rosenblatt, D.A.; Chan, D.; Greenwood, D.C.; Norat, T. Blood concentrations of carotenoids and retinol and lung cancer risk: An update of the WCRF–AICR systematic review of published prospective studies. *Cancer Med.* **2016**, *5*, 2069–2083. [CrossRef]

10. Darvin, M.E.; Sterry, W.; Lademann, J.; Vergou, T. The role of carotenoids in human skin. *Molecules* **2011**, *16*, 10491–10506. [CrossRef]

11. Castenmiller, J.J.M.; West, C.E. Bioavailability and bioconversion of carotenoids. *Annu. Rev. Nutr.* **1998**, *18*, 19–38. [CrossRef]

12. Khoo, H.-E.; Prasad, K.N.; Kong, K.-W.; Jiang, Y.; Ismail, A. Carotenoids and their isomers: Color pigments in fruits and vegetables. *Molecules* **2011**, *16*, 1710–1738. [CrossRef]

13. Alscher, R.G.; Hess, J.L. *Antioxidants in Higher Plants*; CRC Press: Boca Raton, FL, USA, 2017; ISBN 1351369148.

14. Mercadante, A.Z.; Egeland, E.S.; Britton, G.; Liaaen-Jensen, S.; Pfander, H. *Carotenoids Handbook*; Britton, G., Liaaen-Jensen, S., Pfander, H., Eds.; Birkhäuser Basel: Basel, Switzerland, 2004.

15. Chinembiri, T.N.; Du Plessis, L.H.; Gerber, M.; Hamman, J.H.; Du Plessis, J. Review of natural compounds for potential skin cancer treatment. *Molecules* **2014**, *19*, 11679–11721. [CrossRef]

16. Deming, D.M.; Erdman, J.W. Mammalian carotenoid absorption and metabolism. *Pure Appl. Chem.* **2007**, *71*, 2213–2223. [CrossRef]

17. El-Agamey, A.; Lowe, G.M.; McGarvey, D.J.; Mortensen, A.; Phillip, D.M.; Truscott, T.G.; Young, A.J. Carotenoid radical chemistry and antioxidant/pro-oxidant properties. *Arch. Biochem. Biophys.* **2004**, *430*, 37–48. [CrossRef]

18. Rodriguez-Concepcion, M.; Avalos, J.; Bonet, M.L.; Boronat, A.; Gomez-Gomez, L.; Hornero-Mendez, D.; Limon, M.C.; Meléndez-Martínez, A.J.; Olmedilla-Alonso, B.; Palou, A.; et al. A global perspective on carotenoids: Metabolism, biotechnology, and benefits for nutrition and health. *Prog. Lipid Res.* **2018**, *70*, 62–93. [CrossRef]

19. Darvin, M.E.; Fluhr, J.W.; Caspers, P.; van der Pool, A.; Richter, H.; Patzelt, A.; Sterry, W.; Lademann, J. In vivo distribution of carotenoids in different anatomical locations of human skin: Comparative assessment with two different raman spectroscopy methods. *Exp. Dermatol.* **2009**, *18*, 1060–1063. [CrossRef]

20. Lowe, G.M.; Bilton, R.F.; Davies, I.G.; Ford, T.C.; Billington, D.; Young, A.J. Carotenoid composition and antioxidant potential in subfractions of human low-density lipoprotein. *Ann. Clin. Biochem.* **1999**, *36*, 323–332. [CrossRef]

21. During, A.; Dawson, H.D.; Harrison, E.H. Carotenoid transport is decreased and expression of the lipid transporters SR-BI, NPC1L1, and ABCA1 is downregulated in Caco-2 cells treated with ezetimibe. *J. Nutr.* **2005**, *135*, 2305–2312. [CrossRef]

22. Nagao, A. Bioavailability of dietary carotenoids: Intestinal absorption and metabolism. *Japan Agric. Res. Q.* **2014**, *48*, 385–392. [CrossRef]
23. Kong, K.-W.; Khoo, H.-E.; Prasad, K.N.; Ismail, A.; Tan, C.-P.; Rajab, N.F. Revealing the power of the natural red pigment lycopene. *Molecules* **2010**, *15*, 959–987. [CrossRef]
24. Stahl, W.; Sundquist, A.R.; Hanusch, M.; Schwarz, W.; Sies, H. Separation of β-carotene and lycopene geometrical isomers in biological samples. *Clin. Chem.* **1993**, *39*, 810–814.
25. Lademann, J.; Meinke, M.C.; Sterry, W.; Darvin, M.E. Carotenoids in human skin. *Exp. Dermatol.* **2011**, *20*, 377–382. [CrossRef]
26. Khachik, F.; Spangler, C.J.; Smith, J.C.; Canfield, L.M.; Steck, A.; Pfander, H. Identification, Quantification, and Relative Concentrations of Carotenoids and Their Metabolites in Human Milk and Serum. *Anal. Chem.* **1997**, *69*, 1873–1881. [CrossRef]
27. Eroglu, A.; Hruszkewycz, D.P.; Dela Sena, C.; Narayanasamy, S.; Riedl, K.M.; Kopec, R.E.; Schwartz, S.J.; Curley, R.W.; Harrison, E.H. Naturally occurring eccentric cleavage products of provitamin A β-carotene function as antagonists of retinoic acid receptors. *J. Biol. Chem.* **2012**, *287*, 15886–15895. [CrossRef]
28. Jaswir, I.; Noviendri, D.; Hasrini, R.F.; Octavianti, F. Carotenoids: Sources, medicinal properties and their application in food and nutraceutical industry. *J. Med. Plants Res* **2011**, *5*, 7119–7131.
29. Insel, P.M. *Discovering Nutrition*; Jones & Bartlett Publishers: Burlington, MA, USA, 2013; ISBN 1449632947.
30. Bunea, A.; Rugină, D.; Pintea, A.; Andrei, S.; Bunea, C.; Pop, R.; Bele, C. Carotenoid and fatty acid profiles of bilberries and cultivated blueberries from Romania. *Chem. Pap.* **2012**, *66*, 935–939. [CrossRef]
31. Ježek, D.; Tripalo, B.; Brnčić, M.; Karlović, D.; Rimac Brnčić, S.; Vikić-Topić, D.; Karlović, S. Dehydration of celery by infrared drying. *Croat. Chem. Acta* **2008**, *81*, 325–331.
32. Mikulic-Petkovsek, M.; Stampar, F.; Veberic, R.; Sircelj, H. Wild Prunus fruit species as a rich source of bioactive compounds. *J. Food Sci.* **2016**, *81*, C1928–C1937. [CrossRef]
33. Müller-Maatsch, J.; Sprenger, J.; Hempel, J.; Kreiser, F.; Carle, R.; Schweiggert, R.M. Carotenoids from gac fruit aril (Momordica cochinchinensis [Lour.] Spreng.) are more bioaccessible than those from carrot root and tomato fruit. *Food Res. Int.* **2017**, *99*, 928–935. [CrossRef]
34. Radošević, K.; Srček, V.G.; Bubalo, M.C.; Brnčić, S.R.; Takács, K.; Redovniković, I.R. Assessment of glucosinolates, antioxidative and antiproliferative activity of broccoli and collard extracts. *J. Food Compos. Anal.* **2017**, *61*, 59–66. [CrossRef]
35. Davinelli, S.; Nielsen, M.E.; Scapagnini, G. Astaxanthin in skin health, repair, and disease: A comprehensive review. *Nutrients* **2018**, *10*, 522. [CrossRef]
36. Hussein, G.; Sankawa, U.; Goto, H.; Matsumoto, K.; Watanabe, H. Astaxanthin, a carotenoid with potential in human health and nutrition. *J. Nat. Prod.* **2006**, *69*, 443–449. [CrossRef]
37. Camera, E.; Mastrofrancesco, A.; Fabbri, C.; Daubrawa, F.; Picardo, M.; Sies, H.; Stahl, W. Astaxanthin, canthaxanthin and β-carotene differently affect UVA-induced oxidative damage and expression of oxidative stress-responsive enzymes. *Exp. Dermatol.* **2009**, *18*, 222–231. [CrossRef]
38. Ito, N.; Seki, S.; Ueda, F. The protective role of astaxanthin for UV-induced skin deterioration in healthy people—a randomized, double-blind, placebo-controlled trial. *Nutrients* **2018**, *10*, 817. [CrossRef]
39. Ambati, R.R.; Moi, P.S.; Ravi, S.; Aswathanarayana, R.G. Astaxanthin: Sources, extraction, stability, biological activities and its commercial applications - A review. *Mar. Drugs* **2014**, *12*, 128–152. [CrossRef]
40. Van Chuyen, H.; Eun, J.B. Marine carotenoids: Bioactivities and potential benefits to human health. *Crit. Rev. Food Sci. Nutr.* **2017**, *57*, 2600–2610. [CrossRef]
41. Peng, J.; Yuan, J.-P.; Wu, C.-F.; Wang, J.-H. Fucoxanthin, a marine carotenoid present in brown seaweeds and diatoms: Metabolism and bioactivities relevant to human health. *Mar. Drugs* **2011**, *9*, 1806–1828. [CrossRef]
42. Anese, M.; Mirolo, G.; Beraldo, P.; Lippe, G. Effect of ultrasound treatments of tomato pulp on microstructure and lycopene in vitro bioaccessibility. *Food Chem.* **2013**, *136*, 458–463. [CrossRef]
43. Carbonell-Capella, J.M.; Šic Žlabur, J.; Rimac Brnčić, S.; Barba, F.J.; Grimi, N.; Koubaa, M.; Brnčić, M.; Vorobiev, E. Electrotechnologies, microwaves, and ultrasounds combined with binary mixtures of ethanol and water to extract steviol glycosides and antioxidant compounds from Stevia rebaudiana leaves. *J. food Process. Preserv.* **2017**, *41*, e13179. [CrossRef]
44. Darvin, M.; Patzelt, A.; Gehse, S.; Schanzer, S.; Benderoth, C.; Sterry, W.; Lademann, J. Cutaneous concentration of lycopene correlates significantly with the roughness of the skin. *Eur. J. Pharm. Biopharm.* **2008**, *69*, 943–947. [CrossRef] [PubMed]

45. Fiedor, J.; Burda, K. Potential role of carotenoids as antioxidants in human health and disease. *Nutrients* **2014**, *6*, 466–488. [CrossRef] [PubMed]

46. Ribeiro, D.; Freitas, M.; Silva, A.M.S.; Carvalho, F.; Fernandes, E. Antioxidant and pro-oxidant activities of carotenoids and their oxidation products. *Food Chem. Toxicol.* **2018**, *120*, 681–699. [CrossRef] [PubMed]

47. Eichler, O.; Sies, H.; Stahl, W. Divergent Optimum Levels of Lycopene, β-Carotene and Lutein Protecting Against UVB Irradiation in Human Fibroblasts. *Photochem. Photobiol.* **2004**, *75*, 503–506. [CrossRef]

48. Edge, R.; McGarvey, D.J.; Truscott, T.G. The carotenoids as anti-oxidants—a review. *J. Photochem. Photobiol. B Biol.* **1997**, *41*, 189–200. [CrossRef]

49. Edge, R.; Truscott, T. Singlet oxygen and free radical reactions of retinoids and carotenoids—a review. *Antioxidants* **2018**, *7*, 5. [CrossRef]

50. Mayne, S.T.; Cartmel, B.; Scarmo, S.; Lin, H.; Leffell, D.J.; Welch, E.; Ermakov, I.; Bhosale, P.; Bernstein, P.S.; Gellermann, W. Noninvasive assessment of dermal carotenoids as a biomarker of fruit and vegetable intake. *Am. J. Clin. Nutr.* **2010**, *92*, 794–800. [CrossRef] [PubMed]

51. Tsuruoka, H.; Khovidhunkit, W.; Brown, B.E.; Fluhr, J.W.; Elias, P.M.; Feingold, K.R. Scavenger receptor class B type I is expressed in cultured keratinocytes and epidermis. Regulation in response to changes in cholesterol homeostasis and barrier requirements. *J. Biol. Chem.* **2002**, *277*, 2916–2922. [CrossRef]

52. Priyadarshani, A.M.B. Insights of hypercarotenaemia: A brief review. *Clin. Nutr. ESPEN* **2018**, *23*, 19–24. [CrossRef]

53. Maharshak, N.; Shapiro, J.; Trau, H. Carotenoderma–a review of the current literature. *Int. J. Dermatol.* **2003**, *42*, 178–181. [CrossRef]

54. Palombo, P.; Fabrizi, G.; Ruocco, V.; Ruocco, E.; Fluhr, J.; Roberts, R.; Morganti, P. Beneficial long-term effects of combined oral/topical antioxidant treatment with the carotenoids lutein and zeaxanthin on human skin: A double-blind, placebo-controlled study. *Skin Pharmacol. Physiol.* **2007**, *20*, 199–210. [CrossRef] [PubMed]

55. Darvin, M.E.; Fluhr, J.W.; Meinke, M.C.; Zastrow, L.; Sterry, W.; Lademann, J. Topical beta-carotene protects against infra-red-light-induced free radicals. *Exp. Dermatol.* **2011**, *20*, 125–129. [CrossRef] [PubMed]

56. Dreher, F.; Maibach, H. Protective effects of topical antioxidants in humans. *Curr. Probl. Dermatol.* **2001**, *29*, 157–164. [PubMed]

57. Lademann, J.; Caspers, P.J.; Van Der Pol, A.; Richter, H.; Patzelt, A.; Zastrow, L.; Darvin, M.; Sterry, W.; Fluhr, J.W. In vivo Raman spectroscopy detects increased epidermal antioxidative potential with topically applied carotenoids. *Laser Phys. Lett.* **2009**, *6*, 76–79. [CrossRef]

58. Bogdan Allemann, I.; Baumann, L. Antioxidants used in skin care formulations. *Skin Therapy Lett.* **2008**, *13*, 5–9. [PubMed]

59. Rodríguez-Luna, A.; Ávila-Román, J.; González-Rodríguez, M.L.; Cózar, M.J.; Rabasco, A.M.; Motilva, V.; Talero, E. Fucoxanthin-containing cream prevents epidermal hyperplasia and UVB-induced skin erythema in mice. *Mar. Drugs* **2018**, *16*, 378. [CrossRef] [PubMed]

60. Darvin, M.; Zastrow, L.; Sterry, W.; Lademann, J. Effect of supplemented and topically applied antioxidant substances on human tissue. *Skin Pharmacol. Physiol.* **2006**, *19*, 238–247. [CrossRef]

61. Scarmo, S.; Cartmel, B.; Lin, H.; Leffell, D.J.; Welch, E.; Bhosale, P.; Bernstein, P.S.; Mayne, S.T. Significant correlations of dermal total carotenoids and dermal lycopene with their respective plasma levels in healthy adults. *Arch. Biochem. Biophys.* **2010**, *504*, 34–39. [CrossRef]

62. von Lintig, J.; Sies, H.; Mayne, S.T.; Cartmel, B.; Scarmo, S.; Jahns, L.; Ermakov, I.V.; Gellermann, W. Resonance Raman spectroscopic evaluation of skin carotenoids as a biomarker of carotenoid status for human studies. *Arch. Biochem. Biophys.* **2013**, *539*, 163–170.

63. Sies, H.; Stahl, W. Nutritional Protection against Skin Damage from Sunlight. *Annu. Rev. Nutr.* **2004**, *24*, 173–200. [CrossRef]

64. Lademann, J.; Köcher, W.; Yu, R.; Meinke, M.C.; Na Lee, B.; Jung, S.; Sterry, W.; Darvin, M.E. Cutaneous carotenoids: The mirror of lifestyle? *Skin Pharmacol. Physiol.* **2014**, *27*, 201–207. [CrossRef]

65. Darvin, M.E.; Fluhr, J.W.; Schanzer, S.; Richter, H.; Patzelt, A.; Meinke, M.C.; Zastrow, L.; Golz, K.; Doucet, O.; Sterry, W.; et al. Dermal carotenoid level and kinetics after topical and systemic administration of antioxidants: Enrichment strategies in a controlled in vivo study. *J. Dermatol. Sci.* **2011**, *64*, 53–58. [CrossRef]

66. Godar, D.E. UV Doses Worldwide - Invited Review. *Photochem. Photobiol.* **2005**, *81*, 736–749. [CrossRef]

67. Sanches Silveira, J.E.P.; Myaki Pedroso, D.M. UV light and skin aging. *Rev. Environ. Health* **2014**, *29*, 243–254. [CrossRef] [PubMed]

68. Grant, W.B.; Holick, M.F. Benefits and requirements of vitamin D for optimal health: A review. *Altern. Med. Rev.* **2005**, *10*, 94–111.

69. Matsumura, Y.; Ananthaswamy, H.N. Toxic effects of ultraviolet radiation on the skin. *Toxicol. Appl. Pharmacol.* **2004**, *195*, 298–308. [CrossRef]

70. Juzeniene, A.; Moan, J. Beneficial effects of UV radiation other than via vitamin D production. *Dermatoendocrinol.* **2012**, *4*, 109–117. [CrossRef] [PubMed]

71. Bosch, R.; Philips, N.; Suárez-Pérez, J.; Juarranz, A.; Devmurari, A.; Chalensouk-Khaosaat, J.; González, S. Mechanisms of Photoaging and Cutaneous Photocarcinogenesis, and Photoprotective Strategies with Phytochemicals. *Antioxidants* **2015**, *4*, 248–268. [CrossRef]

72. Timares, L.; Katiyar, S.K.; Elmets, C.A. DNA damage, apoptosis and Langerhans cells - Activators of UV-induced immune tolerance. *Photochem. Photobiol.* **2008**, *84*, 422–436. [CrossRef]

73. Biesalski, H.K.; Obermueller-Jevic, U.C. UV light, beta-carotene and human skin - Beneficial and potentially harmful effects. *Arch. Biochem. Biophys.* **2001**, *389*, 1–6. [CrossRef]

74. Césarini, J.P.; Michel, L.; Maurette, J.M.; Adhoute, H.; Béjot, M. Immediate effects of UV radiation on the skin: Modification by an antioxidant complex containing carotenoids. *Photodermatol. Photoimmunol. Photomed.* **2003**, *19*, 182–189. [CrossRef]

75. Randhawa, M.; Seo, I.S.; Liebel, F.; Southall, M.D.; Kollias, N.; Ruvolo, E. Visible light induces melanogenesis in human skin through a photoadaptive response. *PLoS ONE* **2015**, *10*, e0130949. [CrossRef]

76. Liebel, F.; Kaur, S.; Ruvolo, E.; Kollias, N.; Southall, M.D. Irradiation of skin with visible light induces reactive oxygen species and matrix-degrading enzymes. *J. Invest. Dermatol.* **2012**, *132*, 1901–1907. [CrossRef]

77. Mahmoud, B.H.; Hexsel, C.L.; Hamzavi, I.H.; Lim, H.W. Effects of visible light on the skin. *Photochem. Photobiol.* **2008**, *84*, 450–462. [CrossRef]

78. Fernández-García, E. Skin protection against UV light by dietary antioxidants. *Food Funct.* **2014**, *5*, 1994–2003. [CrossRef]

79. Grether-Beck, S.; Marini, A.; Jaenicke, T.; Stahl, W.; Krutmann, J. Molecular evidence that oral supplementation with lycopene or lutein protects human skin against ultraviolet radiation: Results from a double-blinded, placebo-controlled, crossover study. *Br. J. Dermatol.* **2017**, *176*, 1231–1240. [CrossRef]

80. Parrado, C.; Philips, N.; Gilaberte, Y.; Juarranz, A.; González, S. Oral Photoprotection: Effective Agents and Potential Candidates. *Front. Med.* **2018**, *5*, 1–19. [CrossRef] [PubMed]

81. Freitas, J.V.; Junqueira, H.C.; Martins, W.K.; Baptista, M.S.; Gaspar, L.R. Antioxidant role on the protection of melanocytes against visible light-induced photodamage. *Free Radic. Biol. Med.* **2019**, *131*, 399–407. [CrossRef]

82. Fernández-García, E. Photoprotection of human dermal fibroblasts against ultraviolet light by antioxidant combinations present in tomato. *Food Funct.* **2014**, *5*, 285–290. [CrossRef]

83. Stahl, W.; Sies, H. Carotenoids and protection against solar UV radiation. *Skin Pharmacol. Physiol.* **2002**, *15*, 291–296. [CrossRef]

84. Rabinovich, L.; Kazlouskaya, V. Herbal sun protection agents: Human studies. *Clin. Dermatol.* **2018**, *36*, 369–375. [CrossRef]

85. Stahl, W.; Sies, H. Photoprotection by dietary carotenoids: Concept, mechanisms, evidence and future development. *Mol. Nutr. Food Res.* **2012**, *56*, 287–295. [CrossRef]

86. Afaq, F.; Mukhtar, H. Photochemoprevention by botanical antioxidants. *Skin Pharmacol. Appl. Skin Physiol.* **2002**, *15*, 297–306. [CrossRef]

87. Rizwan, M.; Rodriguez-Blanco, I.; Harbottle, A.; Birch-Machin, M.A.; Watson, R.E.B.; Rhodes, L.E. Tomato paste rich in lycopene protects against cutaneous photodamage in humans in vivo: A randomized controlled trial. *Br. J. Dermatol.* **2011**, *164*, 154–162. [CrossRef]

88. Stahl, W.; Sies, H. β-Carotene and other carotenoids in protection from sunlight. *Am. J. Clin. Nutr.* **2012**, *96*, 1179–1184. [CrossRef]

89. Stahl, W.; Krutmann, J. Systemische photoprotektion durch karotinoide. *Hautarzt* **2006**, *57*, 281–285. [CrossRef]

90. Krinsky, N.I.; Landrum, J.T.; Bone, R.A. Biologic mechanisms of the protective role of lutein and zeaxanthin in the eye. *Annu. Rev. Nutr.* **2003**, *23*, 171–201. [CrossRef]

91. Junghans, A.; Sies, H.; Stahl, W. Macular pigments lutein and zeaxanthin as blue light filters studied in liposomes. *Arch. Biochem. Biophys.* **2001**, *391*, 160–164. [CrossRef]

92. Boukari, F.; Jourdan, E.; Fontas, E.; Montaudié, H.; Castela, E.; Lacour, J.-P.; Passeron, T. Prevention of melasma relapses with sunscreen combining protection against UV and short wavelengths of visible light: A prospective randomized comparative trial. *J. Am. Acad. Dermatol.* **2015**, *72*, 189–190. [CrossRef]

93. Juturu, V.; Bowman, J.P.; Deshpande, J. Overall skin tone and skin-lightening-improving effects with oral supplementation of lutein and zeaxanthin isomers: A double-blind, placebo-controlled clinical trial. *Clin. Cosmet. Investig. Dermatol.* **2016**, *9*, 325–332. [CrossRef]

94. Wölfle, U.; Seelinger, G.; Bauer, G.; Meinke, M.C.; Lademann, J.; Schempp, C.M. Reactive molecule species and antioxidative mechanisms in normal skin and skin aging. *Skin Pharmacol. Physiol.* **2014**, *27*, 316–332. [CrossRef]

95. Greul, A.K.; Grundmann, J.U.; Heinrich, F.; Pfitzner, I.; Bernhardt, J.; Ambach, A.; Biesalski, H.K.; Gollnick, H. Photoprotection of UV-irradiated human skin: An antioxidative combination of vitamins E and C, carotenoids, selenium and proanthocyanidins. *Skin Pharmacol. Appl. Skin Physiol.* **2002**, *15*, 307–315. [CrossRef]

96. Wertz, K.; Seifert, N.; Hunziker, P.B.; Riss, G.; Wyss, A.; Lankin, C.; Goralczyk, R. β-carotene inhibits UVA-induced matrix metalloprotease 1 and 10 expression in keratinocytes by a singlet oxygen-dependent mechanism. *Free Radic. Biol. Med.* **2004**, *37*, 654–670. [CrossRef]

97. Wertz, K.; Hunziker, P.B.; Seifert, N.; Riss, G.; Neeb, M.; Steiner, G.; Hunziker, W.; Goralczyk, R. β-carotene interferes with ultraviolet light A-induced gene expression by multiple pathways. *J. Invest. Dermatol.* **2005**, *124*, 428–434. [CrossRef]

98. Desmarchelier, C.; Borel, P. Overview of carotenoid bioavailability determinants: From dietary factors to host genetic variations. *Trends Food Sci. Technol.* **2017**, *69*, 270–280. [CrossRef]

99. Fernández-García, E.; Carvajal-Lérida, I.; Jarén-Galán, M.; Garrido-Fernández, J.; Pérez-Gálvez, A.; Hornero-Méndez, D. Carotenoids bioavailability from foods: From plant pigments to efficient biological activities. *Food Res. Int.* **2012**, *46*, 438–450. [CrossRef]

100. Köpcke, W.; Krutmann, J. Protection from sunburn with β-carotene - A meta-analysis. *Photochem. Photobiol.* **2008**, *84*, 284–288. [CrossRef]

101. Heinrich, U.; Gärtner, C.; Wiebusch, M.; Eichler, O.; Sies, H.; Tronnier, H.; Stahl, W. Supplementation with β-Carotene or a Similar Amount of Mixed Carotenoids Protects Humans from UV-Induced Erythema. *J. Nutr.* **2018**, *133*, 98–101. [CrossRef]

102. Mathews-Roth, M.M.; Pathak, M.A.; Parrish, J.; Fitzpatrick, T.B.; Kass, E.H.; Toda, K.; Clemens, W. A clinical trial of the effects of oral beta-carotene on the responses of human skin to solar radiation. *J. Invest. Dermatol.* **1972**, *59*, 349–353. [CrossRef]

103. Lee, J.; Jiang, S.; Levine, N.; Watson, R.R. Carotenoid supplementation reduces erythema in human skin after simulated solar radiation exposure. *Proc. Soc. Exp. Biol. Med.* **2000**, *223*, 170–174. [CrossRef]

104. Epstein, J.H. Effects of β-carotene on ultraviolet induced cancer formation in the hairless mouse skin. *Photochem. Photobiol.* **1977**, *25*, 211–213. [CrossRef]

105. Santamaria, L.; Bianchi, A.; Arnaboldi, A.; Andreoni, L.; Bermond, P. Dietary carotenoids block photocarcinogenic enhancement by benzo (a) pyrene and inhibit its carcinogenesis in the dark. *Experientia* **1983**, *39*, 1043–1045. [CrossRef]

106. Palozza, P.; Serini, S.; Torsello, A.; Di Nicuolo, F.; Maggiano, N.; Ranelletti, F.O.; Wolf, F.I.; Calviello, G. Mechanism of activation of caspase cascade during β-carotene-induced apoptosis in human tumor cells. *Nutr. Cancer* **2003**, *47*, 76–87. [CrossRef]

107. Guruvayoorappan, C.; Kuttan, G. β-Carotene down-regulates inducible nitric oxide synthase gene expression and induces apoptosis by suppressing bcl-2 expression and activating caspase-3 and p53 genes in B16F-10 melanoma cells. *Nutr. Res.* **2007**, *27*, 336–342. [CrossRef]

108. Frieling, U.M.; Schaumberg, D.A.; Kupper, T.S.; Muntwyler, J.; Hennekens, C.H. A randomized, 12-year primary-prevention trial of beta carotene supplementation for nonmelanoma skin cancer in the physicians' health study. *Arch. Dermatol.* **2000**, *136*, 179–184. [CrossRef]

109. Hennekens, C.H.; Buring, J.E.; Manson, J.E.; Stampfer, M.; Rosner, B.; Cook, N.R.; Belanger, C.; LaMotte, F.; Gaziano, J.M.; Ridker, P.M.; et al. Lack of Effect of Long-Term Supplementation with Beta Carotene on the Incidence of Malignant Neoplasms and Cardiovascular Disease. *N. Engl. J. Med.* **2002**, *334*, 1145–1149. [CrossRef]

110. Greenberg, E.R.; Baron, J.A.; Stukel, T.A.; Stevens, M.M.; Mandel, J.S.; Spencer, S.K.; Elias, P.M.; Lowe, N.; Nierenberg, D.W.; Bayrd, G. A clinical trial of beta carotene to prevent basal-cell and squamous-cell cancers of the skin. *N. Engl. J. Med.* **1990**, *323*, 789–795. [CrossRef]

111. Wertz, K.; Seifert, N.; Buchwald Hunziker, P.; Riss, G.; Wyss, A.; Hunziker, W.; Goralczyk, R. β-Carotene interference with UVA-induced gene expression by multiple pathways. *Pure Appl. Chem.* **2006**, *78*, 1539–1550. [CrossRef]

112. Minami, Y.; Kawabata, K.; Kubo, Y.; Arase, S.; Hirasaka, K.; Nikawa, T.; Bando, N.; Kawai, Y.; Terao, J. Peroxidized cholesterol-induced matrix metalloproteinase-9 activation and its suppression by dietary β-carotene in photoaging of hairless mouse skin. *J. Nutr. Biochem.* **2009**, *20*, 389–398. [CrossRef]

113. Marini, A.; Jaenicke, T.; Grether-Beck, S.; Le Floc'h, C.; Cheniti, A.; Piccardi, N.; Krutmann, J. Prevention of polymorphic light eruption by oral administration of a nutritional supplement containing lycopene, β-carotene, and L actobacillus johnsonii: Results from a randomized, placebo-controlled, double-blinded study. *Photodermatol. Photoimmunol. Photomed.* **2014**, *30*, 189–194. [CrossRef]

114. Mathews Roth, M.M.; Pathak, M.A.; Fitzpatrick, T.B.; Harber, L.H.; Kass, E.H. Beta Carotene Therapy for Erythropoietic Protoporphyria and Other Photosensitivity Diseases. *Arch. Dermatol.* **1977**, *113*, 1229–1232. [CrossRef]

115. Böhm, F.; Edge, R.; Foley, S.; Lange, L.; Truscott, T.G. Antioxidant inhibition of porphyrin-induced cellular phototoxicity. *J. Photochem. Photobiol. B* **2001**, *65*, 177–183. [CrossRef]

116. Harper, P.; Wahlin, S. Treatment options in acute porphyria, porphyria cutanea tarda, and erythropoietic protoporphyria. *Curr. Treat. Options Gastroenterol.* **2007**, *10*, 444–455. [CrossRef]

117. Offord, E.A.; Gautier, J.C.; Avanti, O.; Scaletta, C.; Runge, F.; Krämer, K.; Applegate, L.A. Photoprotective potential of lycopene, β-carotene, vitamin E, vitamin C and carnosic acid in UVA-irradiated human skin fibroblasts. *Free Radic. Biol. Med.* **2002**, *32*, 1293–1303. [CrossRef]

118. Aust, O.; Stahl, W.; Sies, H.; Tronnier, H.; Heinrich, U. Supplementation with tomato-based products increases lycopene, phytofluene, and phytoene levels in human serum and protects against UV-light-induced erythema. *Int. J. Vitam. Nutr. Res.* **2005**, *75*, 54–60. [CrossRef]

119. Stahl, W.; Heinrich, U.; Aust, O.; Tronnier, H.; Sies, H. Lycopene-rich products and dietary photoprotection. *Photochem. Photobiol. Sci.* **2006**, *5*, 238–242. [CrossRef]

120. Cooperstone, J.L.; Tober, K.L.; Riedl, K.M.; Teegarden, M.D.; Cichon, M.J.; Francis, D.M.; Schwartz, S.J.; Oberyszyn, T.M. Tomatoes protect against development of UV-induced keratinocyte carcinoma via metabolomic alterations. *Sci. Rep.* **2017**, *7*, 5106. [CrossRef]

121. Evans, J.A.; Johnson, E.J. Something New Under the Sun: Lutein—s Role in Skin Health. *Am. J. Lifestyle Med.* **2009**, *3*, 349–352. [CrossRef]

122. Ascenso, A.; Ribeiro, H.; Marques, H.C.; Oliveira, H.; Santos, C.; Simões, S. Chemoprevention of photocarcinogenesis by lycopene. *Exp. Dermatol.* **2014**, *23*, 874–878. [CrossRef]

123. Heinrich, U.; Tronnier, H.; Stahl, W.; Béjot, M.; Maurette, J.M. Antioxidant supplements improve parameters related to skin structure in humans. *Skin Pharmacol. Physiol.* **2006**, *19*, 224–231. [CrossRef]

124. O'Connor, I.; O'Brien, N. Modulation of UVA light-induced oxidative stress by β-carotene, lutein and astaxanthin in cultured fibroblasts. *J. Dermatol. Sci.* **1998**, *16*, 226–230. [CrossRef]

125. Rao, A.R.; Sindhuja, H.N.; Dharmesh, S.M.; Sankar, K.U.; Sarada, R.; Ravishankar, G.A. Effective inhibition of skin cancer, tyrosinase, and antioxidative properties by astaxanthin and astaxanthin esters from the green alga Haematococcus pluvialis. *J. Agric. Food Chem.* **2013**, *61*, 3842–3851. [CrossRef]

126. Komatsu, T.; Sasaki, S.; Manabe, Y.; Hirata, T.; Sugawara, T. Preventive effect of dietary astaxanthin on UVA-induced skin photoaging in hairless mice. *PLoS ONE* **2017**, *12*, 1–16. [CrossRef]

127. Tanaka, T.; Shnimizu, M.; Moriwaki, H. Cancer chemoprevention by carotenoids. *Molecules* **2012**, *17*, 3202–3242. [CrossRef]

128. Hix, L.M.; Lockwood, S.F.; Bertram, J.S. Upregulation of connexin 43 protein expression and increased gap junctional communication by water soluble disodium disuccinate astaxanthin derivatives. *Cancer Lett.* **2004**, *211*, 25–37. [CrossRef]

129. Scarmo, S.N. *Noninvasive Measurement of Carotenoids in Human Skin as a Biomarker of Fruit and Vegetable Intake*; Yale University: New Haven, CA, USA, 2009; ISBN 1109588186.

130. Jyonouchi, H.; Sun, S.; Iijima, K.; Gross, M.D. Antitumor activity of astaxanthin and its mode of action. *Nutr. Cancer* **2000**, *36*, 59–65. [CrossRef]

131. Tominaga, K.; Hongo, N.; Karato, M.; Yamashita, E. Cosmetic benefits of astaxanthin on humans subjects. *Acta Biochim. Pol.* **2012**, *59*, 43–47. [CrossRef]

132. Tominaga, K.; Hongo, N.; Fujishita, M.; Takahashi, Y.; Adachi, Y. Protective effects of astaxanthin on skin deterioration. *J. Clin. Biochem. Nutr.* **2017**, *61*, 33–39. [CrossRef]

133. Yoon, H.-S.; Cho, H.H.; Cho, S.; Lee, S.-R.; Shin, M.-H.; Chung, J.H. Supplementing with Dietary Astaxanthin Combined with Collagen Hydrolysate Improves Facial Elasticity and Decreases Matrix Metalloproteinase-1 and -12 Expression: A Comparative Study with Placebo. *J. Med. Food* **2014**, *17*, 810–816. [CrossRef]

134. Suganuma, K.; Shiobara, M.; Sato, Y.; Nakanuma, C.; Maekawa, T.; Ohtsuki, M.; Yazawa, K.; Imokawa, G. Anti-aging and functional improvement effects for the skin by functional foods intakes: Clinical effects on skin by oral ingestion of preparations containing Astaxanthin and Vitamins C and E. *Jichi Med. Univ. J.* **2012**, *35*, 25–33.

135. Suganuma, K.; Nakajima, H.; Ohtsuki, M.; Imokawa, G. Astaxanthin attenuates the UVA-induced up-regulation of matrix-metalloproteinase-1 and skin fibroblast elastase in human dermal fibroblasts. *J. Dermatol. Sci.* **2010**, *58*, 136–142. [CrossRef]

136. Chalyk, N.E.; Klochkov, V.A.; Bandaletova, T.Y.; Kyle, N.H.; Petyaev, I.M. Continuous astaxanthin intake reduces oxidative stress and reverses age-related morphological changes of residual skin surface components in middle-aged volunteers. *Nutr. Res.* **2017**, *48*, 40–48. [CrossRef]

137. Singh, K.N.; Patil, S.; Barkate, H. Protective effects of astaxanthin on skin: Recent scientific evidence, possible mechanisms, and potential indications. *J. Cosmet. Dermatol.* **2019**, 1–6. [CrossRef]

138. Park, J.H.; Yeo, I.J.; Han, J.H.; Suh, J.W.; Lee, H.P.; Hong, J.T. Anti-inflammatory effect of astaxanthin in phthalic anhydride-induced atopic dermatitis animal model. *Exp. Dermatol.* **2018**, *27*, 378–385. [CrossRef]

139. Meephansan, J.; Rungjang, A.; Yingmema, W.; Deenonpoe, R.; Ponnikorn, S. Effect of astaxanthin on cutaneous wound healing. *Clin. Cosmet. Investig. Dermatol.* **2017**, *10*, 259–265. [CrossRef]

140. González, S.; Astner, S.; An, W.; Pathak, M.A.; Goukassian, D. Dietary lutein/zeaxanthin decreases ultraviolet B-induced epidermal hyperproliferation and acute inflammation in hairless mice. *J. Invest. Dermatol.* **2003**, *121*, 399–405. [CrossRef]

141. Astner, S.; Wu, A.; Chen, J.; Philips, N.; Rius-Diaz, F.; Parrado, C.; Mihm, M.C.; Goukassian, D.A.; Pathak, M.A.; González, S. Dietary lutein/zeaxanthin partially reduces photoaging and photocarcinogenesis in chronically UVB-irradiated Skh-1 hairless mice. *Skin Pharmacol. Physiol.* **2007**, *20*, 283–291. [CrossRef]

142. Heinen, M.M.; Hughes, M.C.; Ibiebele, T.I.; Marks, G.C.; Green, A.C.; van der Pols, J.C. Intake of antioxidant nutrients and the risk of skin cancer. *Eur. J. Cancer* **2007**, *43*, 2707–2716. [CrossRef]

143. Philips, N.; Keller, T.; Hendrix, C.; Hamilton, S.; Arena, R.; Tuason, M.; Gonzalez, S. Regulation of the extracellular matrix remodeling by lutein in dermal fibroblasts, melanoma cells, and ultraviolet radiation exposed fibroblasts. *Arch. Dermatol. Res.* **2007**, *299*, 373–379. [CrossRef]

144. Schwartz, S.; Frank, E.; Gierhart, D.; Simpson, P.; Frumento, R. Zeaxanthin-based dietary supplement and topical serum improve hydration and reduce wrinkle count in female subjects. *J. Cosmet. Dermatol.* **2016**, *15*, e13–e20. [CrossRef]

145. Di Mascio, P.; Kaiser, S.; Sies, H. Lycopene as the most efficient biological carotenoid singlet oxygen quencher. *Arch. Biochem. Biophys.* **1989**, *274*, 532–538. [CrossRef]

146. Tinkler, J.H.; Böhm, F.; Schalch, W.; Truscott, T.G. Dietary carotenoids protect human cells from damage. *J. Photochem. Photobiol. B Biol.* **1994**, *26*, 283–285. [CrossRef]

147. Pirayesh Islamian, J.; Mehrali, H. Lycopene as a carotenoid provides radioprotectant and antioxidant effects by quenching radiation-induced free radical singlet oxygen: An overview. *Cell J.* **2015**, *16*, 386–391.

148. Böhm, F.; Edge, R.; Burke, M.; Truscott, T.G. Dietary uptake of lycopene protects human cells from singlet oxygen and nitrogen dioxide–ROS components from cigarette smoke. *J. Photochem. Photobiol. B Biol.* **2001**, *64*, 176–178. [CrossRef]

149. Stahl, W.; Heinrich, U.; Wiseman, S.; Eichler, O.; Sies, H.; Tronnier, H. Dietary Tomato Paste Protects against Ultraviolet Light–Induced Erythema in Humans. *J. Nutr.* **2018**, *131*, 1449–1451. [CrossRef]

150. Böhm, F.; Edge, R.; Truscott, G. Interactions of dietary carotenoids with activated (singlet) oxygen and free radicals: Potential effects for human health. *Mol. Nutr. Food Res.* **2012**, *56*, 205–216. [CrossRef]

151. McArdle, F.; Rhodes, L.E.; Parslew, R.A.; Close, G.L.; Jack, C.I.; Friedmann, P.S.; Jackson, M.J. Effects of oral vitamin E and β-carotene supplementation on ultraviolet radiation-induced oxidative stress in human skin. *Am. J. Clin. Nutr.* **2004**, *80*, 1270–1275. [CrossRef]

152. White, A.L.; Jahnke, L.S. Contrasting effects of UV-A and UV-B on photosynthesis and photoprotection of β-carotene in two Dunaliella spp. *Plant Cell Physiol.* **2002**, *43*, 877–884. [CrossRef]

153. Schagen, S.K.; Zampeli, V.A.; Makrantonaki, E.; Zouboulis, C.C. Discovering the link between nutrition and skin aging. *Dermatoendocrinol.* **2012**, *4*, 37–41. [CrossRef]

154. Pryor, W.A.; Stahl, W.; Rock, C.L. Beta Carotene: From Biochemistry to Clinical Trials. *Nutr. Rev.* **2009**, *58*, 39–53. [CrossRef]

155. Bayerl, C. Beta-carotene in dermatology: Does it help? *Acta Dermatovenerol. Alp. Pannonica Adriat.* **2008**, *17*, 160–162,164–166.

156. Guerin, M.; Huntley, M.E.; Olaizola, M. Haematococcus astaxanthin: Applications for human health and nutrition. *Trends Biotechnol.* **2003**, *21*, 210–216. [CrossRef]

157. Sofiah, A.S.; Lesmana, R.; Farenia, R.; Adi, S. Astaxanthin cream alters type I procollagen and Matrix metalloproteinase-1 (MMP-1) gene expression induced by ultraviolet B irradiation in rat skin. *J. Biomed. Clin. Sci.* **2018**, *3*, 62–67.

158. Heo, S.J.; Jeon, Y.J. Protective effect of fucoxanthin isolated from Sargassum siliquastrum on UV-B induced cell damage. *J. Photochem. Photobiol. Biol. B* **2009**, *95*, 101–107. [CrossRef]

159. Matsui, M.; Tanaka, K.; Higashiguchi, N.; Okawa, H.; Yamada, Y.; Tanaka, K.; Taira, S.; Aoyama, T.; Takanishi, M.; Natsume, C.; et al. Protective and therapeutic effects of fucoxanthin against sunburn caused by UV irradiation. *J. Pharmacol. Sci.* **2016**, *132*, 55–64. [CrossRef]

160. Shimoda, H.; Tanaka, J.; Shan, S.; Maoka, T. Anti-pigmentary activity of fucoxanthin and its influence on skin mRNA expression of melanogenic molecules. *J. Pharm. Pharmacol.* **2010**, *62*, 1137–1145. [CrossRef]

161. Al-Delaimy, W.K.; van Kappel, A.L.; Ferrari, P.; Slimani, N.; Steghens, J.P.; Bingham, S.; Johansson, I.; Wallström, P.; Overvad, K.; Tjønneland, A.; et al. Plasma levels of six carotenoids in nine European countries: Report from the European Prospective Investigation into Cancer and Nutrition (EPIC). *Public Health Nutr.* **2004**, *7*, 713–722. [CrossRef]

162. Al-Delaimy, W.K.; Ferrari, P.; Slimani, N.; Pala, V.; Johansson, I.; Nilsson, S.; Mattisson, I.; Wirfalt, E.; Galasso, R.; Palli, D.; et al. Plasma carotenoids as biomarkers of intake of fruits and vegetables: Individual-level correlations in the European Prospective Investigation into Cancer and Nutrition (EPIC). *Eur. J. Clin. Nutr.* **2005**, *59*, 1397–1408. [CrossRef]

163. Meinke, M.C.; Müller, R.; Bechtel, A.; Haag, S.F.; Darvin, M.E.; Lohan, S.B.; Ismaeel, F.; Lademann, J. Evaluation of carotenoids and reactive oxygen species in human skin after UV irradiation: A critical comparison between in vivo and ex vivo investigations. *Exp. Dermatol.* **2015**, *24*, 194–197. [CrossRef]

164. Meinke, M.C.; Darvin, M.E.; Vollert, H.; Lademann, J. Bioavailability of natural carotenoids in human skin compared to blood. *Eur. J. Pharm. Biopharm.* **2010**, *76*, 269–274. [CrossRef]

165. Segger, D.; Schönlau, F. Supplementation with Evelle®improves skin smoothness and elasticity in a double-blind, placebo-controlled study with 62 women. *J. Dermatolog. Treat.* **2004**, *15*, 222–226. [CrossRef]

166. Meinke, M.C.; Nowbary, C.K.; Schanzer, S.; Vollert, H.; Lademann, J.; Darvin, M.E. Influences of orally taken carotenoid-rich curly kale extract on collagen I/elastin index of the skin. *Nutrients* **2017**, *9*, 775. [CrossRef]

167. Cho, S.; Lee, D.H.; Won, C.-H.; Kim, S.M.; Lee, S.; Lee, M.-J.; Chung, J.H. Differential effects of low-dose and high-dose beta-carotene supplementation on the signs of photoaging and type I procollagen gene expression in human skin in vivo. *Dermatology* **2010**, *221*, 160–171. [CrossRef]

168. Pandel, R.; Poljšak, B.; Godic, A.; Dahmane, R. Skin photoaging and the role of antioxidants in its prevention. *ISRN Dermatol.* **2013**, *930164*. [CrossRef]

169. Berthon, J.Y.; Nachat-Kappes, R.; Bey, M.; Cadoret, J.-P.; Renimel, I.; Filaire, E. Marine algae as attractive source to skin care. *Free Radic. Res.* **2017**, *51*, 555–567. [CrossRef]

170. Urikura, I.; Sugawara, T.; Hirata, T. Protective effect of fucoxanthin against UVB-induced skin photoaging in hairless mice. *Biosci. Biotechnol. Biochem.* **2011**, *75*, 757–760. [CrossRef]

171. Kishimoto, Y.; Tani, M.; Uto-Kondo, H.; Iizuka, M.; Saita, E.; Sone, H.; Kurata, H.; Kondo, K. Astaxanthin suppresses scavenger receptor expression and matrix metalloproteinase activity in macrophages. *Eur. J. Nutr.* **2010**, *49*, 119–126. [CrossRef]

172. Eren, B.; Tuncay Tanrıverdi, S.; Aydın Köse, F.; Özer, Ö. Antioxidant properties evaluation of topical astaxanthin formulations as anti-aging products. *J. Cosmet. Dermatol.* **2019**, *18*, 242–250. [CrossRef]

173. Kindlund, P.J. Astaxanthin. *Nutrafoods* **2011**, *10*, 27–31. [CrossRef]
174. Daubrawa, F.; Sies, H.; Stahl, W. Astaxanthin diminishes gap junctional intercellular communication in primary human fibroblasts. *J. Nutr.* **2005**, *135*, 2507–2511. [CrossRef]

antioxidants

MDPI

Article

In Vitro Digestion of Human Milk: Influence of the Lactation Stage on the Micellar Carotenoids Content

Ana A. O. Xavier [1], Juan E. Garrido-López [1], Josefa Aguayo-Maldonado [2],
Juan Garrido-Fernández [1], Javier Fontecha [3] and Antonio Pérez-Gálvez [1,*]

[1] Food Phytochemistry Department, Instituto de la Grasa (CSIC), Campus Universitario, Building 46,
    41013 Sevilla, Spain
[2] Unidad de Neonatología, Hospital Virgen del Rocío, 41013 Sevilla, Spain
[3] Institute of Food Science Research (CSIC-UAM), 28049 Madrid, Spain
* Correspondence: aperez@ig.csic.es; Tel.: +34-954-611550

Received: 29 June 2019; Accepted: 6 August 2019; Published: 7 August 2019

Abstract: Human milk is a complex fluid with nutritive and non-nutritive functions specifically structured to cover the needs of the newborn. The present study started with the study of carotenoid composition during progress of lactation (colostrum, collected at 3–5 d postpartum; mature milk, collected at 30 d postpartum) with samples donated from full-term lactating mothers (women with no chronic diseases, nonsmokers on a regular diet without supplements, $n = 30$). Subsequently, we applied an in vitro protocol to determine the micellarization efficiency of the carotenoids, which were separated by HPLC and quantified by the external standard method. That in vitro protocol is tailored for the biochemistry of the digestive tract of a newborn. To the best of our knowledge, the present study is the first report of carotenoids micellar contents, obtained in vitro. This study reveals, from the in vitro perspective, that colostrum and mature milk produce significant micellar contents of carotenoids despite lipids in milk are within highly complex structures. Indeed, the lactation period develops some influence on the micellarization efficiency, influence that might be attributed to the dynamics of the milk fat globule membrane (MFGM) during the progress of lactation.

Keywords: breastfeeding; newborn; human colostrum; carotenes and xanthophylls; in vitro digestibility; micellar lipids

1. Introduction

Carotenoids are one of the five families of natural pigments widely distributed in the vegetal kingdom, in photosynthetic microorganisms, and some fungi. These pigments are isoprenoid compounds, which contribute to the light-harvesting process and filter harmful light radiations, and display antioxidant activity. According to their structural features, carotenoids are classified as carotenes (pure hydrocarbons) or xanthophylls (oxygenated carotenes), that is, the classical C40 carotenoids, while apo-carotenoids which arise from carotenoid metabolism that shortens the C40 structure, and C30 carotenoids that been described in bacteria complete the main subfamilies of carotenoids. For a more detailed lecture about structural features and biosynthesis of carotenoids, some general overviews are suggested [1,2]. It is the lipophilic nature of these compounds the main feature that marks the mode of interaction with other biomolecules, and the environment where these processes take place. Hence, carotenoids are biosynthesized in fruit and vegetal tissues in the corresponding plastid organelles, which present their own membrane, and behave in a lipophilic environment with other biomolecules. Therefore, the tissues where carotenoids may incorporate and develop further activities should resemble in a similar way either a membrane macrostructure or a lipophilic surrounding.

Additionally, the carotenoids significantly contribute to the nutritional value of natural sources where they distribute, e.g., fruit and vegetables, algae, eggs, and fish [2]. Animals exclusively rely on diet to incorporate these compounds to both internal tissues and systemic circulation where they perform significant functions and biological activities. Hence, almost 10% of the described carotenoids in nature present the structural requisites to transform into vitamin A [3], while all carotenoids exert important functions in immunity, participate in the antioxidant defense system, and are related to a reduced risk of developing chronic diseases [4,5]. These biological activities are the fundamentals to provide evidence of the inverse association between the ingestion of carotenoid-containing fruits and vegetables, or serum carotenoid levels, with risk for various chronic diseases [6–12].

Taking both topics into account, the study of bioavailability of carotenoids (from digestion to bioactivity) is the key piece of knowledge to establish the actual contribution of carotenoids to human health, from the provitamin A value, with specific significance in some developing countries where the dietary supply of vitamin A sources if often limited [13] to the other biological activities. The considerable research literature regarding the measurement of carotenoids appearing in serum after the ingestion of different carotenoid-rich sources (fruit and vegetables, juices, oils …) either in human beings or animals, was the starting point to learn that efficiency of the bioaccessibility of carotenoids is controlled by several factors that may limit or enhance their final disposition in systemic circulation and inner tissues. Thus, the disseminated work of Castenmiller and West [14] systematically reviewed the factors that influence carotenoid bioaccessibility and bioactivity in vivo, which were condensed in the mnemotechnic term *SLAMENGHI* where each letter represents one variable. It is necessary to remark again that the influence of these factors was initially ascertained from in vivo experiments, while our more recent advancements have made use of in vitro techniques to determine the considerable number of factors that affect bioaccessibility with a similar confidence to the in vivo approach.

Milk is a multifaceted fluid with nutritive and non-nutritive functions and actions, comprising a wide range of molecules in solution, colloidal aggregates, and complex structures such as the milk fat globules. Even cells and microorganisms are observed in this somehow 'alive' secretion that impacts the physiology of the newborn at different tissues and at different levels (digestive tract, microbiota, immune system, vision, and cognitive development) [15]. Indeed, breast milk lactation provides both the mother and the infant with significant benefits already documented [16–18]. Regarding the quantitative composition, the lipophilic content shows the highest variability so far. Hence, the lipid nutrients represent ca. 45–55% of the total caloric intake, while among lipophilic micro-nutrients counted in human milk, vitamin A is in the higher concentration range (200–600 mg/L) that in addition to carotenoids overcome the concentration of other liposoluble vitamins [19,20]. This is an issue of interest when we realize that the lipophilic vitamins in the fetus are considerably low as the accumulation in the fetus through the placenta during the accretion period is not effective, and the fetal liver stores of lipophilic vitamins are limited [21]. This means that most of the newborn present a deficit of vitamin A after delivery, a deficit that is amended with the lactation. Accordingly, most of the infants are subjected to certain oxidative stress because of the transition from a situation of hypoxia in the uterus (partial O_2 pressure at 25–35 mm Hg) to the extrauterine environment with partial O_2 pressure at 100 mm Hg. In addition, the onset of the mitochondrial respiration initiates the concomitant production of reactive oxygen species [22], whereas the immaturity of the endogenous antioxidant system means an additional tendency to increased oxidative stress after delivery [23]. Consequently, there should be a significant interest in determining the bioavailability of those human milk compounds that contribute to relief the oxidative stress of the infant [24]. In vitro digestion protocols have been developed to measure the extent of digestibility, assimilation, and first-pass metabolism of food components, a very active research subject in recent decades [25]. Thus, the digestion protocols have been applied to measure both the bioaccessibility of carotenoids from natural food sources (Figure 1) and the impact of factors in the efficiency of that process [26–31]. Indeed, in vitro digestion protocols have been specifically developed for mimicking the gastric conditions of the newborn [32] to determine the bioaccessibility of lipolysis and proteolysis from human, cow, and goat milk; infant formula; and a functional beverage based in whey protein concentrate [33–37].

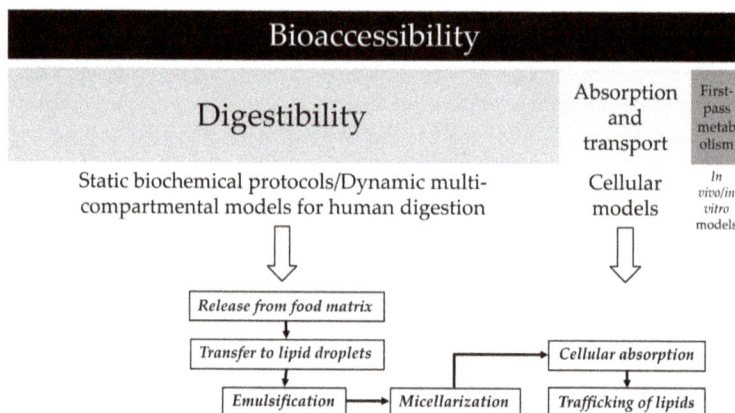

Figure 1. The different contribution of the three sequential steps that yield the bioaccessibility value of carotenoids, including the in vitro models required to measure their efficiency, showing the main molecular events associated to each stage.

It is hypothesized that the application of an in vitro protocol that tailors the experimental conditions to those of the gastrointestinal maturity rate of the newborn might provide with significant data regarding the efficiency of the digestion of carotenoids. Hence, the aim of this study was to measure the micellar contents of carotenoids in human milk at two stages of the lactation period: colostrum and mature milk. The application of an in vitro digestion protocol that follows the harmonized international agreement is the core to allow future comparisons among different pools of samples (pre-term, full-term human milk, infant formula). Therefore, the advance in knowledge regarding the accumulation of lipids with significant contributions to the health of the neonate will expand from the state-of-the-art in vitro techniques to the in vivo arena.

2. Materials and Methods

2.1. Subjects

The study population comprised 30 healthy women, which were recruited within a 12-month period between 2018 and 2019, from a 1200–1500 pregnant population that delivers at the hospital area in Sevilla (Spain). They gave birth to healthy full-term neonates (37–40 weeks). Eligible participants in this study were non-smoking mothers with no chronic disease. In addition, mothers following any special diets, vegetarians, or those taking supplements were not included. Exclusion criteria applied were pathologies and/or infections during the gestation, developmental anomalies in the fetus, or death of the child. Considering our previous studies, 30 volunteers per group would provide enough statistical power to detect (if any) significant differences between colostrum and mature milk of the micellar carotenoid profile [37].

2.2. Reagents

Pepsin from porcine gastric mucosa, porcine bile extract, and pancreatin and lipase from porcine pancreas were obtained from Sigma (St. Louis, MO, USA). Solvents (HPLC-grade) were provided by Romyl (Teknokroma, Barcelona, Spain), and the purified water was obtained from a Milli-Q water purification system (Millipore, Milford, MA, USA).

2.3. Measures

2.3.1. Milk Collection

The samples were collected at the Unidad de Neonatología of the Hospital Universitario Virgen del Rocío (Sevilla, Spain). Full-term mothers donated colostrum at 3–5 days postpartum or mature milk at 30 days postpartum. Milk samples were obtained by collection of the total milk volume of one breast during one milk expression session into a polypropylene bottle. The samples were transported directly to the laboratory and stored at 4 °C. Subsequent analyses of the samples (carotenoid extraction and in vitro digestion) were performed within 1–2 days after collection.

2.3.2. In Vitro Digestion of Human Milk

The experimental conditions described by Ménard et al. [32] were applied with slight modifications. The in vitro protocol does not include the oral phase, while the gastric and intestinal steps mimics the digestive conditions of full-term infants. Colostrum or mature milk sample (6 mL) was mixed with 3.5 mL of gastric fluid and incubated for 60 min under magnetic stirring in a water bath at 37 °C. The composition of the gastric fluid was 94 mM NaCl and 13 mM KCl at pH 5.3 (HCl 1 M). Enzymes of the gastric fluid were pepsin (268 U/mL) and lipase (19 U/mL). At the end of the gastric phase, sample was cooled in water, pH adjusted to 6.6 and mixed with 5.5 mL of intestinal fluid, and the resulting cocktail incubated at 37 °C with magnetic stirring for 60 min. The intestinal fluid was composed of 164 mM NaCl, 10 mM KCl, 85 mM NaHCO$_3$, and bile extract at 3.1 mM bile salt concentration at pH 7. Porcine pancreatin, with 90 U/mL of lipase activity, was added to the intestinal fluid. The upper micellar fraction [38,39] was isolated from digested sample by centrifugation (12,000× g, 5 min, 4 °C) in an Avanti^{TM} J-25 centrifuge (Beckman Coulter^{TM}, Brea, CA, USA) equipped with a Beckman model JA-25.50 rotor (Kildare, Ireland). The micellar fraction was collected and used for measurement of the micellar carotenoid content. Three replicates of the in vitro digestion procedure for each sample (colostrum or mature milk) were carried out.

2.3.3. Extraction of the Carotenoid Fraction

The experimental conditions previously described by Ríos et al. [37] were slightly modified for extraction of carotenoids from human milk. Sample (3 mL) was mixed with 3 mL of KOH:methanol (20% w/v), and the mixture was incubated for 1 h, at 25 °C. After hydrolysis, 6 mL of methanol were added, and the mixture was vortex-mixed for 2 min and cooled at −20 °C for 20 min. Subsequently, the cooled mixture was centrifuged at 10,000× g and 4 °C for 5 min, discarding the upper layer. Diethyl ether (5 mL) and hexane (2 mL) were added to the pellet and vortex-mixed for 2 min. Then, 5 mL of NaCl 10% (w/v) was added, and the sample was vortex-mixed again for 2 min. After centrifugation (10,000× g at 4 °C for 5 min), the organic layer was washed with water until neutral pH was reached. The organic extract was evaporated to dryness in a rotatory evaporator at 25 °C, and the residue was dissolved in 0.25 mL of acetone. Carotenoids from micelles were extracted following the same procedure. The final extracts were filtered through a 0.22 μm filter and stored at −20 °C until analysis by HPLC, which was performed within 1 week. The extraction and HPLC analysis of carotenoids was performed under diminished light and avoiding excessive contact with air.

2.3.4. Quantification of Carotenoids in the Carotenoid Extracts

Carotenoid extracts from human milk samples and from micellar fraction were analyzed using a Jasco HPLC (Easton, PA, USA) equipped with quaternary pump (model PU-2089-plus), autosampler (model AS-2055-plus), and diode array detector (MD-2010-plus). Chromatographic data were acquired and managed using the Jasco ChromPass Chromatography Data System software (version 1.8.6.1). Carotenoids were separated on a Luna (Phenomenex, Torrance, CA, USA) C18 column (250 × 4 mm, 5 μm particle size), which was stabilized at 25 °C, using a linear gradient of acetone/water, from 75:25 (v/v) to 95:5 in 5 min, hold 95:5 for 7 min and to 100:0 in 3 min, maintaining this proportion for 10 min, and going back to 75:25 in 5 min. Flow rate was set at 1.5 mL/min and 100 μL of sample was injected. The UV–visible absorption spectra were acquired between 200 and 600 nm and the chromatograms processed at 450 nm. The carotenoids were identified according to elution order on C18 column and characteristics of UV–visible spectrum (λmax, spectral fine structure (% III/II), and peak *cis* (if present) intensity (% AB/AII)), as compared to standards and data available in the literature [40]. Stock solutions were prepared for β-carotene, β-cryptoxanthin, lutein, and lycopene at a concentration of approximately 25 mg/L. Once the exact concentration was determined, working stock solutions for external calibration curves were prepared at 5 concentration levels ranging 0.150–10 mg/L. Lutein and zeaxanthin, as well as their *cis* isomers were quantified as a single peak, while α- and β-carotene, and their *cis* isomers, were quantified as a single peak

2.3.5. Determination of the Lipid Content

The lipid content of human milk samples was determined according to the solvent extraction procedure and then by gravimetry [41].

2.4. Ethics Approval

Participants provided informed consent for inclusion before they participated in the study, which was conducted in accordance with the Declaration of Helsinki. The study protocol was approved by the Ethics Committee of the Hospital Universitario Virgen del Rocío (AGL2017-87884-R).

2.5. Statistical Analysis and Calculation

Efficiency of the micellarization of carotenoids (%) was determined as the ratio of individual carotenoids in micelles to their corresponding content in the milk samples. Due to the non-normality of the distribution of the content of individual carotenoids (Shapiro–Wilk test, $p < 0.05$), the data were analyzed using a non-parametric statistical procedure in the SPSS software (IBM® SPSS® Statistics version 24, IBM, New York, NY, USA). Hence, data are reported as the median, including 25th and 75th percentiles. Considering the sample size and the heteroscedasticity of the variances, we applied the Friedman test to establish whether the carotenoid contents within each group of samples, both in the human milk and in the micellar fraction, were significantly different or not. Then, we applied the Wilcoxon test to analyze which carotenoids contents were significantly different within each group of samples. To analyze whether the micellar contents were significantly different between colostrum and mature milk, we applied the Kruskal–Wallis test. Subsequently, the Mann–Whitney test was applied to analyze the differences among the micellar contents of colostrum and mature milk samples, and to compare the micellar contents within each type of human milk samples. The significance test was set at $p < 0.05$.

3. Results

Figure 2 shows a chromatogram trace of the carotenoids extracted from the micellar fraction obtained after the in vitro digestion of a human mature milk sample. Carotenoids in human milk samples followed the already established trend for the transition from colostrum to mature milk (Table 1). A high content in colostrum for all the carotenoids is significantly reduced in the mature milk, which represents the 13% of the initial colostrum content. This sharp decline is not correlated with a reduction of the fat content as colostrum and mature milk showed total milk fat content in the same range (30–40 mg/mL). Regarding the amounts of the individual carotenoids, lycopene was the carotenoid with the lowest presence in the analyzed samples at both lactation stages, then α+β-carotene (plus *cis* isomers), β-cryptoxanthin and finally the sum of zeaxanthin and lutein (plus *cis* isomers). Xanthophylls (X) and carotenes (C) were equally distributed in colostrum, while mature milk showed a X to C ratio higher than 1.

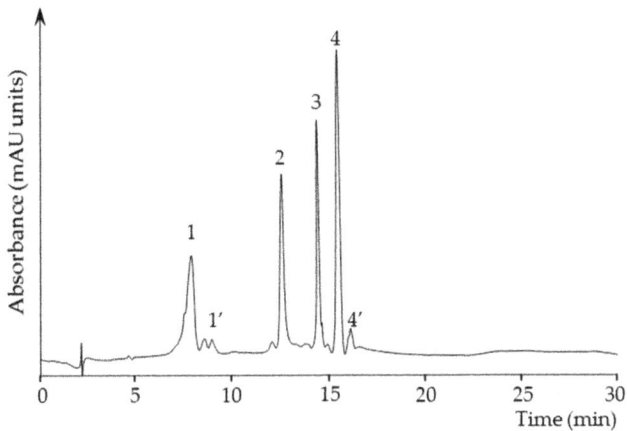

Figure 2. HPLC trace at 450 nm of the carotenoids extracted from the micellar fraction of a human mature milk sample digested in vitro. Carotenoid identification is as follows: 1, zeaxanthin + lutein; 1', *cis*-isomers of zeaxanthin + lutein; 2, β-cryptoxanthin; 3, lycopene; 4, α+β-carotene; 4', *cis*-isomers of α+β-carotene.

Figure 3 depicts the micellar contents of carotenoids in the digested samples of colostrum and mature milk. Regarding the micellar contents in colostrum samples, no significant differences were observed within the X, or within the C. However, when the micellar content of both groups of carotenoids are compared, significant differences were observed, with the X reaching a micellar efficiency of ca. 50%, and below 40% for the C. In the case of the micellar contents of carotenoids after the in vitro digestion of mature milk, no significant differences were observed neither within the X, and within the C as well as in the comparison of the micellar content of X with those of C, reaching both groups of carotenoids a micellar efficiency at the 25–30% range. Finally, the micellar efficiency is significantly higher in colostrum for zeaxanthin + lutein, and β-cryptoxanthin, in comparison with the corresponding micellar contents in mature milk, while lycopene and α+β-carotene are equally micellarized independently of the lactation state.

Table 1. Carotenoid content in colostrum and mature human milk ($n = 30$). Data are expressed in ng/mL.

Carotenoid	25th Percentile	Median [1]	75th Percentile
		Colostrum [2]	
Zeaxanthin + lutein [3]	190	349	664
β-cryptoxanthin	195	285	464
lycopene	114	224	371
α+β-carotene [4]	133	241	265
		Mature Milk [5]	
Zeaxanthin + lutein [3]	26.7	43.9	67.0
β-cryptoxanthin	9.51	37.6	62.7
lycopene	6.79	11.6	19.8
α+β-carotene [4]	6.45	41.8	51.5

[1] Data are significantly different when colostrum and mature milk values are compared (Mann–Whitney test, $p < 0.01$). [2] Data were not significantly different (Friedman test, $p > 0.05$). [3,4] Data include the *cis* isomers when observed. [5] Data were not significantly different (Friedman test, $p > 0.05$) except for lycopene.

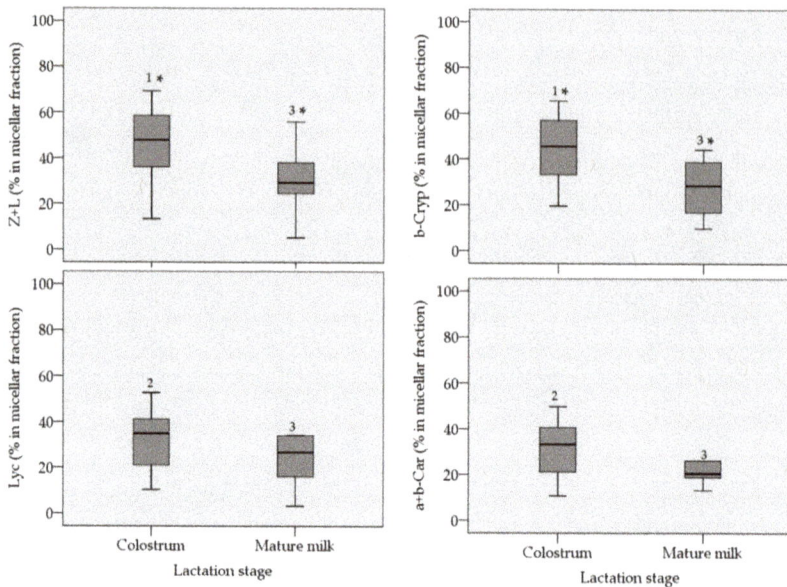

Figure 3. Micellar carotenoid contents (median percentage values ± 75th percentile) in colostrum and mature milk. Data were analyzed using non-parametric statistical procedure. Data denoted with the same number were not significantly different, both for colostrum and mature milk (Wilcoxon test, $p > 0.05$). Data denoted with an asterisk were significantly different (Mann–Whitney test, $p < 0.05$).

4. Discussion

As new data are added to literature regarding the micellarization efficiency of carotenoids from natural food sources, a sort of database is built that allows to compare the efficiency of that process in different foods, and how it is influenced by factors such as the food structure, food matrix and processing features, as well as by those physiological issues that might be reproduced with the in vitro digestion protocol [25]. Thus, according to the results provided with this study, human colostrum yields micellar carotenoid contents like those provided by whole and semi-skimmed milk and yogurt (45%) enriched with a water-soluble lutein formulation [39], and higher than the micellar carotenoid content observed after the in vitro digestion of a puree of cooked vegetables (29–37%) containing spinach, carrot, and tomato [42]. Indeed, the values observed in this study are close to those reported (58–78%) for the in vitro digestion of cooked durum wheat pasta, and cooked pasta containing eggs [27]. It should be noted that although the in vitro protocols applied in the cited studies present slight modifications among them, the experimental approach is basically the same [42]. However, the data presented in the study of Lipkie et al. [43] for micellarization efficiency of carotenoids from mature human milk and infant formula samples are significantly lower than those presented in this study. Although the authors made slight modifications to reduce the sample volume in digesta and reproduce the biochemical gut conditions, the observed differences demonstrate a need for a consensus protocol, so that interlaboratory comparisons are feasible and reproducible.

Therefore, we applied an in vitro model that tailors the experimental conditions of the digestion to those of an infant [32], although the international consensus already published is generally followed [44]. In that protocol for 'infants', the ratio to meal value is higher than in the 'adult' protocol, while the enzymatic capability contained in the digestive fluids is considerably lower. Consequently, the comparison of the data presented in this study with those published in the literature is not straightforward. Certainly, it could be assumed that efficiency of carotenoid micellarization from colostrum and mature milk is higher than it should be, as the sequence of physicochemical processes involved in the digestion and absorption of lipids (Figure 1) is highly dependent of the action of digestive secretions.

Another issue that should be discussed when presenting data acquired with in vitro techniques is their correlation with in vivo data, if available, or their predictive power to the real in vivo scenario. In addition, that comparison among in vitro and in vivo results is useful to reinforce the application of the in vitro protocol or to introduce modifications to obtain a better correlation. There are some in vivo data regarding the assimilation of carotenoids from human milk and infant formula. Thus, the serum lutein content of breastmilk-fed infants increased from the baseline level after delivery to reach almost six-fold higher serum lutein values than in formula-fed infants [45]. The same study showed that to obtain similar serum lutein levels as in the breastmilk-fed group, the formula-fed group required a supplementation in the formula with 4-times more lutein than the content observed in the breastmilk. This study denoted the efficiency of the digestive system of the infants to assimilate carotenoids (lipids) from the structured milk fat globules, and that at equal lutein concentration in the fed, the efficiency is higher from human milk. Our data agreed with this efficiency as it was expressed above. Efficiency regarding the micellarization process of colostrum, which for lutein and zeaxanthin is in the highest range when compared with other carotenoid food sources. Application of the in vitro digestion protocol to infant formula would complete the vision of the correlation of in vitro with the in vivo data. Another study showed that plasma carotenoids of infants fed with breastmilk are in the same concentration range, independently of whether they belong to the X or to the C groups [46]. Thus, β-cryptoxanthin, lycopene, and β-carotene in the plasma of breastfed children was ca. 20 μg/L at 2–6 weeks after delivery. Although the authors stated that the data present statistical differences in concentration, our results point to a similar accessibility of carotenoids from mature milk (Figure 3), e.g., they potentially contribute equally to the pool of plasma carotenoids.

Regarding the change of behavior observed in the lactation stage from a higher in vitro micellarization of X in colostrum to the same micellar contents for X and C after digestion of mature milk, this issue could be related with changes in the lipid and protein composition of the MFGM during the lactation period. Lipids from the mammary gland tissue are secreted in milk as milk fat globules enveloped by a three-layered membrane, the MFGM arising from the cellular membrane of the mammary epithelium [47]. That macrostructure, which contains a wide range of polar lipids and membrane associated proteins, envelopes a core rich in triacylglycerides and lipophilic micronutrients. Therefore, the activity of enzymatic secretions may take different rates to access the core of triacylglycerides where most of the carotenoids are accumulated, depending on the composition of the MFGM. The dynamics of the composition of the MFGM has been demonstrated over the course of lactation for the protein profiles [48], while the evolution of the lipids has not been fully characterized so far [49]. Consequently, the influence of the composition of the milk fat globule membrane in the extent of lipolysis is a feasible hypothesis that deserves further attention. One limitation of this study is the lack of in vivo data regarding the bioaccessibility of carotenoids in the newborn to correlate with the in vitro results, which would additionally support the experimental conditions applied for digestion of the colostrum and mature human milk samples.

5. Conclusions

To the best of our knowledge, the present study is the first report of carotenoids micellar contents in both colostrum and mature human milk after the application of an in vitro protocol, which is tailored for the biochemistry of the digestive tract of a newborn. Our study reveals, from the in vitro perspective, that colostrum and mature milk produce significant micellar contents of carotenoids despite lipids in milk being within highly complex structures. Indeed, the lactation period develops some influence in the micellarization efficiency, influence that might be attributed to the dynamics of the MFGM during the progress of lactation. These results may serve for comparative aims for future studies regarding digestibility of human milk and the influence of other factors far from the lactation period. This new knowledge is a challenge for food industry that could be achieved with the experimental work to develop in this proposal. The design of infant formula for specific infant sub-populations may apply the experimental protocol outlined here to test the efficiency of the digestibility of the supplemented nutrients included in the formulation. Indeed, our study encourages closely mimicking, as far as possible, the architecture of the complex lipid structures presented in human milk, which are lacked in infant formula, considering the digestive efficiency they present.

Author Contributions: Formal analysis, J.E.G.-L. and J.G.-F.; Methodology, A.A.O.X., J.E.G.-L. and J.G.-F.; Data curation, A.A.O.X. and A.P.-G.; Conceptualization, J.F. and A.P.-G.; Investigation, J.A.-M., J.F. and A.P.-G.; Supervision, J.A.-M.; Writing—original draft, J.F. and A.P.-G.

Funding: This research was funded by the Ministerio de Ciencia, Investigación y Universidades, Agencia Estatal de Investigación, grant number AGL2017-87884-R. A.A.O.X. is a fellow of the Science without Borders Program of the Brazilian National Council for Scientific and Technological Development, fellowship 238163/2012-1.

Acknowledgments: The authors are grateful to the lactating mothers who donate milk at the Banco de Leche Materna Donada, Hospital Virgen del Rocío de Sevilla.

Conflicts of Interest: The authors declare no conflict of interest.

References

1. Britton, G. General overview of carotenoid biosynthesis. In *Carotenoid Esters in Foods: Physical, Chemical and Biological Properties*, 1st ed.; Mercadante, A.Z., Ed.; The Royal Society of Chemistry: London, UK, 2019; pp. 109–136.

2. Liaaen-Jensen, S. Basic Carotenoid Chemistry. In *Carotenoids in Health and Disease*, 1st ed.; Krinsky, N.I., Mayne, S.T., Sies, H., Eds.; Marcel Dekker: New York, NY, USA, 2004; pp. 1–30.

3. Bauernfeind, J.C. Carotenoid vitamin A precursors and analogs in foods and feeds. *J. Agric. Food Chem.* **1972**, *20*, 456–473. [CrossRef] [PubMed]

4. Bendich, A. Non-provitamin A activity of carotenoids: Immunoenhancement. *Trends Food Sci. Technol.* **1991**, *2*, 127–130. [CrossRef]

5. Stahl, W.; Sies, H. Bioactivity and protective effects of natural carotenoids. *Biochim. Biophys. Acta* **2005**, *1740*, 101–107. [CrossRef] [PubMed]

6. Tapiero, H.; Townsend, D.M.; Tew, K.D. The role of carotenoids in the prevention of human pathologies. *Biomed. Pharmacother.* **2004**, *58*, 100–110. [CrossRef] [PubMed]

7. Arathi, B.P.; Sowmya, P.R.R.; Vijay, K.; Baskaran, V.; Lakshminarayana, R. Metabolomics of carotenoids: The challenges and prospects. A review. *Trends Food Sci. Technol.* **2015**, *45*, 105–117. [CrossRef]

8. Cooper, D.A. Carotenoids in health and disease: Recent evaluations, research recommendations and the consumer. *J. Nutr.* **2004**, *134*, 221S–224S. [CrossRef] [PubMed]

9. Giovannucci, E.; Ascherio, A.; Rimm, E.B.; Stampfer, M.J.; Colditz, G.A.; Willett, W.C. Intake of carotenoids and retinol in relation to risk of prostate cancer. *J. Natl. Cancer Inst.* **1995**, *87*, 1767–1776. [CrossRef] [PubMed]

10. Hu, F.; Wang, Y.B.; Zhang, W.; Liang, J.; Lin, C.; Li, D.; Wang, F.; Pang, D.; Zhao, Y. Carotenoids and breast cancer risk: A meta-analysis and meta regression. *Breast Cancer Res. Treat.* **2012**, *131*, 239–253. [CrossRef]

11. World Cancer Research Fund. Continuous Update Project. Cancer Prevention Recommendations. Available online: https://wcrf.org/int/research-we-fund/continuous-update-project-cup (accessed on 25 June 2019).

12. Wang, Y.; Chung, S.J.; McCullough, M.L.; Song, W.O.; Fernandez, M.L.; Koo, S.I.; Chun, O.K. Dietary carotenoids are associated with cardiovascular disease risk biomarkers mediated by serum carotenoid concentrations. *J. Nutr.* **2014**, *144*, 1067–1074. [CrossRef]

13. Stoltzfus, R.J.; Hakima, M.; Miller, K.W.; Rasmussen, K.M.; Dawiesah, S.I.; Habicht, J.P.; Dibley, M.J. High dose vitamin A supplementation of breast-feeding Indonesian mothers: Effects on the vitamin A status of mother and infant. *J. Nutr.* **1993**, *123*, 666–675. [CrossRef]

14. Castenmiller, J.J.M.; West, C.E. Bioavailability and bioconversion of carotenoids. *Annu. Rev. Nutr.* **1998**, *18*, 19–38. [CrossRef] [PubMed]

15. Walker, A. Breast milk as the gold standard for protective nutrients. *J. Pediatr.* **2010**, *156*, S3–S7. [CrossRef] [PubMed]

16. American Academy of Pediatrics. Breastfeeding and the use of human milk. American Academy of pediatrics policy statement. *Pediatrics* **2005**, *115*, 496–506. [CrossRef] [PubMed]

17. Horta, B.L.; Bahl, R.; Martines, J.C.; Victora, C.G. *Evidence on the Long-Term Effects of Breastfeeding. Systematic Reviews and Meta-Analyses*; World Health Organization: Geneva, Switzerland, 2007; ISBN 978 92 4 159523 0.

18. Ip, S.; Chung, M.; Raman, G.; Chew, P.; Magula, N.; Devine, D.; Trikalinos, T.; Lau, J. *Breastfeeding and Maternal and Infant Health Outcomes in Developed Countries*; Evidence Report/Technology Assessment No. 153; Agency for Healthcare Research and Quality: Rockville, MD, USA, 2007.

19. Picciano, M.F. Nutrient composition of human milk. *Pediatr. Clin. N. Am.* **2001**, *48*, 53–67. [CrossRef]

20. Ballard, O.; Morrow, A. Human milk composition: Nutrients and bioactive factors. *Pediatr. Clin. N. Am.* **2013**, *60*, 49–74. [CrossRef]

21. Picciano, M.F. Pregnancy and lactation: Physiological adjustments, nutritional requirements and the role of dietary supplemnts. *J. Nutr.* **2003**, *133*, 1997S–2002S. [CrossRef]

22. Singer, D.; Muhlfeld, C. Perinatal adaptation in mammals: The impact of metabolic rate. *Comp. Biochem. Physiol. A Mol. Integr. Physiol.* **2007**, *148*, 780–784. [CrossRef]

23. Franco, M.C.; Kawamoto, E.M.; Gorjao, R.; Rastelli, V.M.; Curi, R.; Scavone, C.; Sawaya, A.L.; Fortes, Z.B.; Sesso, R. Biomarkers of oxidative stress and antioxidant status in children born small for gestational age: Evidence of lipid peroxidation. *Pediatr. Res.* **2007**, *62*, 204–208. [CrossRef]

24. Pérez-Gálvez, A.; Jarén-Galán, M.; Garrido-Fernández, J.; Calvo, M.V.; Visioli, F.; Fontecha, J. Activities, bioavailability, and metabolism of lipids from structural membranes and oils: Promising research on mild cognitive impairment. *Pharmacol. Res.* **2008**, *134*, 299–304. [CrossRef]

25. Pérez-Gálvez, A. In vitro digestion protocols: The benchmark for estimation of in vivo data. In *Carotenoid Esters in Foods: Physical, Chemical and Biological Properties*, 1st ed.; Mercadante, A.Z., Ed.; The Royal Society of Chemistry: London, UK, 2019; pp. 421–458.

26. Reboul, E.; Richelle, M.; Perrot, E.; Desmoulins-Malezet, C.; Pirisi, V.; Borel, P. Bioaccessibility of carotenoids and vitamin E form their main dietary sources. *J. Agric. Food Chem.* **2006**, *54*, 8749–8755. [CrossRef]

27. Werner, S.; Böhm, V. Bioaccessibility of carotenoids and vitamin E from pasta: Evaluation of an in vitro digestion model. *J. Agric. Food Chem.* **2011**, *59*, 1163–1170. [CrossRef] [PubMed]

28. Carbonell-Capella, J.M.; Buniowska, M.; Barba, F.J.; Esteve, M.J.; Frígola, A. Analytical methods for determining bioavailability and bioaccessibility of bioactive compounds from fruits and vegetables: A review. *Compr. Rev. Food Sci. Food Saf.* **2014**, *13*, 155–171. [CrossRef]

29. Rodrigues, D.B.; Mariutti, L.R.B.; Mercadante, A.Z. An in vitro digestion method adapted for carotenoids and carotenoid esters: Moving forward towards standardization. *Food Funct.* **2016**, *7*, 4992–5001. [CrossRef] [PubMed]

30. Estévez-Santiago, R.; Olmedilla-Alonso, B.; Fernández-Jalao, I. Bioaccessibility of provitamin A carotenoids from fruits: Application of a standardised static in vitro digestion method. *Food Funct.* **2016**, *7*, 1354–1366. [CrossRef] [PubMed]

31. Schweiggert, R.M.; Carle, R. Carotenoid deposition in plant and animal foods and its impact on bioavailability. *Crit. Rev. Food Sci. Nutr.* **2017**, *57*, 1807–1830. [CrossRef] [PubMed]

32. Ménard, O.; Bourlieu, C.; De Oliveira, S.C.; Dellarosa, N.; Laghi, L.; Carrière, F.; Capozzi, F.; Dupont, D.; Deglaire, A. A first step towards a consensus static in vitro model for simulating full-term infant digestion. *Food Chem.* **2018**, *240*, 338–345. [CrossRef]

33. Ye, A.; Cui, J.; Singh, H. Effect of fat globule membrane on in vitro digestion of milk fat globules with pancreatic lipase. *Int. Dairy. J.* **2010**, *20*, 822–829. [CrossRef]

34. Berton, A.; Rouvellac, S.; Robert, B.; Rousseau, F.; Lopez, C.; Crenon, I. Effect of the size and interface composition of milk fat globules on their in vitro digestion by the human pancreatic lipase: Native versus homogenized fat globules. *Food Hydrocoll.* **2012**, *29*, 124–134. [CrossRef]

35. Hernández-Alvarez, E.; Blanco-Navarro, I.; Pérez-Sacristán, B.; Sánchez-Siles, L.M.; Granado-Lorencio, F. In vitro digestion-assisted development of a β-cryptoxanthin-rich functional beverage; in vivo validation using systemic response and faecal content. *Food Chem.* **2016**, *208*, 18–25. [CrossRef]

36. Hodgkinson, A.J.; Wallace, O.A.M.; Boggs, I.; Broadhurst, M.; Prosser, C.G. Gastric digestion of cow and goat milk: Impact of infant and young child in vitro digestion conditions. *Food Chem.* **2018**, *245*, 275–281. [CrossRef]

37. Ríos, J.J.; Xavier, A.A.O.; Díaz-Salido, E.; Arenilla-Vélez, I.; Jarén-Galán, M.; Garrido-Fernández, J.; Aguayo-Maldonado, J.; Pérez-Gálvez, A. Xanthophyll esters are found in human colostrum. *Mol. Nutr. Food Res.* **2017**, *61*, 1700296. [CrossRef]

38. Garrett, D.A.; Failla, M.L.; Sarama, R.J. Development of an in vitro digestion method to assess carotenoid bioavailability from meals. *J. Agric. Food Chem.* **1999**, *47*, 4301–4309. [CrossRef]

39. Xavier, A.A.O.; Mercadante, A.Z.; Garrido-Fernández, J.; Pérez-Gálvez, A. Fat content affects bioaccessibility and efficiency of enzymatic hydrolysis of lutein esters added to milk and yogurt. *Food Res. Int.* **2014**, *65*, 171–176. [CrossRef]

40. Britton, G. UV/visible spectroscopy. In *Carotenoids: Spectroscopy*, 1st ed.; Britton, G., Liaaen-Jensen, S., Pfander, H., Eds.; Birkhauser: Basel, Switzerland, 1995; Volume 1B, pp. 13–62.

41. Jensen, R.G.; Lammi-Keefe, C.J.; Koletzko, B. Representative sampling of human milk and the extraction of fat for analysis of environmental lipophilic contaminants. *Toxicol. Environ. Chem.* **1997**, *62*, 229–247. [CrossRef]

42. Garrett, D.A.; Failla, M.L.; Sarama, R.J. Estimation of carotenoid bioavailability from fresh stir-fried vegetables using an in vitro digestion/Caco-2 cell culture model. *J. Nutr. Biochem.* **2000**, *11*, 574–580. [CrossRef]

43. Lipkie, T.E.; Banavara, D.; Shah, B.; Morrow, A.L.; McMahon, R.J.; Jouni, Z.E.; Ferruzzi, M.G. Caco-2 accumulation of lutein is greater from human milk than from infant formula despite similar bioaccessibility. *Mol. Nutr. Food Res.* **2014**, *58*, 2014–2022. [CrossRef]

44. Minekus, M.; Alminger, M.; Alvito, P.; Ballance, S.; Bohn, T.; Bourlieu, C.; Carrière, F.; Boutrou, R.; Corredig, M.; Dupont, D.; et al. A standardized static in vitro digestion method suitable for food—An international consensus. *Food Funct.* **2014**, *5*, 1113–1124. [CrossRef]

45. Bettler, J.; Zimmer, J.P.; Neuringer, M.; DeRusso, P.A. Serum lutein concentrations in healthy term infants fed human milk or infant formula with lutein. *Eur. J. Nutr.* **2010**, *49*, 45–51. [CrossRef]

46. Sommerburg, O.; Meissner, K.; Nelle, M.; Lenhartz, H.; Leichsenring, M. Carotenoid supply in breast-fed and formula-fed neonates. *Eur. J. Pediatr.* **2000**, *159*, 86–90. [CrossRef]

47. McManaman, J.L.; Neville, M.C. Mammary physiology and milk secretion. *Adv. Drug. Deliv. Rev.* **2003**, *55*, 629–641. [CrossRef]

48. Reinhardt, T.A.; Lippolis, J.D. Developmental changes in the milk fat globule membrane proteome during the transition from colostrum to milk. *J. Dairy Sci.* **2008**, *91*, 2307–2318. [CrossRef]

49. Abrahamse, E.; Minekus, M.; van Aken, G.A.; van de Heijning, B.; Knol, J.; Bartke, N.; Oozeer, R.; van der Beek, E.M.; Ludwig, T. Development of the digestive system-experimental challenges and approaches of infant lipid digestion. *Food Dig.* **2012**, *3*, 63–77. [CrossRef] [PubMed]

antioxidants

MDPI

Article

Inclusion Complexes of Lycopene and β-Cyclodextrin: Preparation, Characterization, Stability and Antioxidant Activity

Haixiang Wang [1,2], Shaofeng Wang [1], Hua Zhu [1], Suilou Wang [1] and Jiudong Xing [1,3,*]

[1] Department of Food Quality and Safety, School of Engineering, China Pharmaceutical University, Jiangning District, Nanjing 211198, China
[2] Beijing Advanced Innovation Center for Food Nutrition and Human Health, Beijing Technology and Business University (BTBU), Haidian District, Beijing 100048, China
[3] Pharmaceutical Experimental Training Center, School of Pharmacy, China Pharmaceutical University, Jiangning District, Nanjing 211198, China
*   Correspondence: 1019830736@cpu.edu.cn; Tel.: +86-25-86185754

Received: 30 June 2019; Accepted: 14 August 2019; Published: 16 August 2019

Abstract: In this study, the inclusion complexes of lycopene with β-cyclodextrin (β-CD) were prepared by the precipitation method. Then the inclusion complexes were characterized by the scanning electron microscopy (SEM), ultraviolet-visible spectroscopy (UV), microscopic observation, liquid chromatography, differential scanning calorimetry (DSC) and phase-solubility study. Moreover, the stability and antioxidant activity were tested. The results showed that lycopene was embedded into the cavity of β-CD with a 1:1 stoichiometry. Moreover, the thermal and irradiant stabilities of lycopene were all significantly increased by the formation of lycopene/β-CD inclusion complexes. Antioxidant properties of lycopene and its inclusion complexes were evaluated on the basis of measuring the scavenging activity for 1,1-diphenyl-2-picrylhydrazyl (DPPH), hydroxyl and superoxide anion radicals. The results showed that the scavenging activity of DPPH radicals was obviously increased by the formation of the inclusion complex with β-cyclodextrin at concentrations of 5–30 µg/mL, however, some significant positive effects on the scavenging activity of hydroxyl and superoxide anion radicals were not observed and the reasons are worth further study.

Keywords: lycopene; β-cyclodextrin; inclusion complexes; stability; antioxidant activity

---

1. Introduction

Lycopene, a carotenoid, is an unsaturated lipophilic isoprenoid pigment, which imparts red color to some vegetables and fruits such as tomato, watermelon, pink guava and pink grapefruit [1–3]. It has gained great interest due to its biological properties in the antioxidant activity, anti-inflammatory, cancer prevention and cardiovascular protection [4–9]. Lycopene was widely applied in the food, cosmetics and pharmaceutical industries [10,11]. Ciriminna et al. (2016) had summarized three points of lycopene's emerging applications in these fields: One product was as a nutritional supplement associated with the health benefits (e.g., lycopene soft capsules); another product was Lycosome produced by lycopene micelles embed a whey protein isolate; finally, natural lycopene was used in cosmetic products such as age-defying treatments, facial moisturizers and eye creams [12]. However, the molecular structure of lycopene has several unsaturated bonds, which makes it very unstable and susceptible to light, heat and certain chemical conditions [13–15]. So its application had been seriously limited.

Some possible technologies had been already used to improve the stability of lycopene. For example, Pérez-Masiá et al. (2015) prepared a micro/nanocapsules of lycopene through electrospraying and spray drying. The results showed that the capsules could protect lycopene against thermal degradation [16]. Rocha-Selmi et al. (2013) used gelatin and gum Arabic as the encapsulating agents to prepare microcapsule of lycopene by complex coacervation. The encapsulation efficiency values were above 90% and the degradations of lycopene were decreased in the microcapsule form as compared to its free form [17]. The encapsulation of lycopene with lecithin and α-tocopherol was carried out by a supercritical anti-solvent process. The encapsulated lycopene had better stability than free lycopene. The degradations of lycopene in the encapsulated particles were less than 10% for 28-days when stored under 4 °C [18]. The stability of nanoencapsulated lycopene prepared by interfacial deposition of preformed poly(ε-caprolactone) (PCL) during photosensitization (5 °C–25 °C), heating (60 °C–80 °C) and refrigeration (5 °C) was studied by Pereira dos Santos et al. (2016) and the results showed that nanoencapsulation improved the stability of lycopene under different processing conditions [19]. Bou et al. (2011) investigated the lycopene stability in oil-in-water emulsions, the results showed that lycopene oxidation could be significantly affected by adding free radical scavengers [20]. From previous studies, it could be seen that encapsulation was a common used technology to improve the lycopene's stability.

In the process of encapsulation, cyclodextrins (CDs) are most frequently used materials as the encapsulating agents. Cyclodextrins (CDs) are ring molecules with a hollow cylindrical structure and they have a hydrophobic internal cavity and a hydrophilic external surface. Due to the special structure of CDs, they can improve the stability and water solubility of guest molecules by prepared inclusion complexes and have been applied widely in pharmaceutical, biotechnology and the food industry [21–25]. CDs, such as α-CDs, are widely used to prepare inclusion complexes in some previous reports. For example, the encapsulation in α-cyclodextrins (α-CDs) of tomato oleoresins extracted by supercritical carbon dioxide were prepared by Durante et al. (2016) as ready-to-mix ingredients for novel functional food formulation. α-CD encapsulation had not improved the stability of lycopene [26]. Solvent-free lycopene-rich oleoresins extracted from gac, tomato and watermelon ripe-fruits by supercritical CO_2 were used to obtain stable aqueous suspensions through oleoresin clathration into α-cyclodextrins (α-CDs) and the effects of each lycopene-containing suspension on the viability of human lung adenocarcinoma cells were investigated [27]. Solvent-free oils from the ripe pumpkin extracted by supercritical carbon dioxide as a ready-to-mix oil/α-cyclodextrins (α-CDs) powder were obtained to produce durum wheat pasta. Oil chlatration increased the stability of some bioactives during pasta production and ameliorated poor textural and sensory characteristics of the cooked spaghetti compared with the oil sample [28].

The application of CDs for molecular encapsulation in foods offers several advantages as they are non-toxic, inexpensive, thermally stable, and not hygroscopic. Moreover, they are not absorbed in the upper gastrointestinal tract, and are completely metabolized by the colon microflora [22,29]. CDs mainly include α-, β-, γ-CD and their derivatives, of which β-CD is the most commonly used host for the preparation of inclusion complexes, because its cavity size, at a diameter of 6.0 Å–6.5 Å and a volume of 265 $Å^3$, is suitable for common molecules with molecular weights between 200 g/mol and 800 g/mol [30,31].

In this study, lycopene and β-CD inclusion complexes were prepared and characterized by the scanning electron microscopy (SEM), ultraviolet-visible spectroscopy (UV), microscopic observation, high performance liquid chromatography (HPLC), differential scanning calorimetry (DSC) and phase-solubility study. In addition, the stability and antioxidant activity were also tested. This research provided an effective way to reduce the loss of lycopene in the preservation and improve the utilization of lycopene.

2. Material and Methods

2.1. Material

Lycopene (98% purity) was purchased from the NanJing JingZhu Bio-Technology Co., Ltd. (Nanjing, China). Acetonitrile, Methanol, Ethanol, n-Hexane, Acetone, Salicylic acid, Hydrogen peroxide (H_2O_2), Iron (II) sulfate heptahydrate ($FeSO_4 \cdot 7H_2O$), Pyrogallic acid, Tris(hydroxymethyl)aminomethane (Tris), Hydrochloric acid (HCl) and β-Cyclodextrin (β-CD) were purchased from Sinopharm Chemical Reagent Co., Ltd. (Shanghai, China). 1,1-diphenyl-2-picrylhydrazyl radical (DPPH, ≥97.0% purity) was provided by Phygene Life Science Co., Ltd. (Fuzhou, China). Acetonitrile and methanol were HPLC grade. Other reagents were analytical reagent grade. All experiments were carried out using purified water.

2.2. Preparation of Lycopene/β-CD Inclusion Complexes

The inclusion complexes of lycopene and β-CD were prepared by the co-precipitation method as described in the references [32–34]. According to the solubility curve, approximately 5.0 g of β-CD were dissolved in 100 mL of purified water maintained at 50 °C on a hot plate to make a saturated solution. Approximately 2.4 g of lycopene were then slowly added to the warm β-CD solution to make the molar ratio of lycopene:β-CD = 1:1. After that, the mixtures were stirred for 20 h at 50 °C and later refrigerated overnight at 4 °C. The cold precipitated lycopene/β-CD inclusion complexes were recovered by vacuum filtration and then dried in a convection oven at 50 °C for 24 h. Finally, the complexes were stored in a desiccator at 25 °C until used.

2.3. Characterization of Inclusion Complexes

2.3.1. Entrapment Efficiency (EE)

Entrapment efficiency (EE) was determined according to a reported method [24]. A certain concentration of the lycopene solution was analyzed by the UV-Vis spectrophotometer (TU-1901, Beijing Puxi Instrument CO., Ltd., Beijing, China) and the maximum absorption wavelength of lycopene was 474 nm. The amount of lycopene entrapped in the inclusion complexes was determined as follows: 20 mg of the sample was weighed accurately and washed with 20 mL of diethyl ether anhydrous to remove lycopene on the surface of complexes, then the sample was dispersed in 10 mL of the acetone and n-hexane mixed solution (v:v = 1:1). After ultrasonic extraction at 100 W for 30 min, and finally centrifugation at 4000 rpm for 20 min, the supernatant was obtained and immediately detected by the spectrophotometer at 474 nm. The entrapment efficiency was calculated using the following equation:

$$EE\% = \left(\frac{\text{Amount of entrapped lycopene}}{\text{Total amount of lycopene}} \right) \times 100\% \tag{1}$$

2.3.2. Scanning Electron Microscopy (SEM)

The surface morphology of lycopene, β-CD, lycopene/β-CD inclusion complexes and their physical mixtures were examined by the scanning electron microscope (SEM, SU8020, Hitachi, Japan). The powders were previously fixed on a brass stub using a double-sided adhesive tape and then were made electrically conductive by coating, in a vacuum with a thin layer of gold for 60 s. The pictures were taken at an excitation voltage of 20 kV and a magnification of 4000×.

2.3.3. Microscopic Observation

The microscopic observation was performed by the optical microscope (Eclipse 80i, Nikon, Japan). The pictures of lycopene, β-CD, lycopene/β-CD inclusion complexes and their physical mixtures were recorded at a magnification of 10 × 20.

2.3.4. Ultraviolet-Visible Spectroscopy (UV)

UV spectra were recorded for lycopene, β-CD, lycopene/β-CD inclusion complexes and their physical mixtures using UV spectrophotometer (TU-1901, Beijing Puxi Instrument CO., Ltd., Beijing, China). The measurements were done in the wavelength range from 190 nm to 800 nm and the spectral bandwidth used was 0.5 nm.

2.3.5. High Performance Liquid Chromatography

The HPLC (LC-20A, Shimadzu, Japan) system equipped with a UV/Vis detector was used to characterize the inclusion complexes. The HPLC analytical conditions were achieved on a Ultimate LP-C18 (150 mm × 4.6 mm, 5 μm) column (Welch, Shanghai, China) with a mobile phase containing methanol and acetonitrile in the ratio (90:10, *v/v*) flowing at a rate of 1.0 mL min^{-1}.The column temperature was maintained at 30 °C. The wavelength was monitored at 474 nm and each injection volume was 20 μL.

2.3.6. Differential Scanning Calorimetry (DSC)

The thermal behaviors of lycopene, β-CD, lycopene/β-CD inclusion complexes and their physical mixtures were performed by a differential scanning calorimetry (Q2000, TA Instruments, New Castle, DE, USA). The samples were sealed in aluminum pans and heated at the rate of 10 °C/min from 10 °C to 300 °C in the nitrogen atmosphere.

2.4. Phase-Solubility Study

According to the reported methods [32,35], a series of β-CD solutions were prepared (0, 2, 4, 6, 8, 10 mM). Then an excessive number of lycopene was added to each solution, and ultrasonic treatment was carried out subsequently for 30 min. Following that, the obtained solutions were stirred for 24 h. After that, all the suspensions were filtered through a 0.45 μm syringe filter and analyzed by the UV spectrophotometer at 474 nm.

2.5. Stability Experiments

The thermal stability was carried out by keeping the solution of lycopene and lycopene/β-CD at 50 °C for different time (0, 15, 30, 45, 60, 90, 120, 150, 210 min), and the absorbance was subsequently recorded. Simultaneously, the photostability of the solution was investigated. The solution was irradiated under a lamp for 48 h and the absorbance was also recorded.

2.6. Antioxidant Activity

2.6.1. Measurement of DPPH Radical Scavenging Activity

The DPPH radical scavenging activity was measured according to the method reported by Mishra et al. (2012) with some modification [36]. 2.0 mL of each sample solution was added to 2.0 mL of 0.06 mM DPPH in ethanol. After gentle mixing, the mixture was left to stand for 30 min in the dark, and the absorbance was measured at 517 nm. The DPPH radical scavenging activity was calculated according to the following equation:

$$\text{DPPH radical scavenging activity } (\%) = [1 - (A_1 - A_3)/A_2] \times 100\% \tag{2}$$

where A_1 was the absorbance in the presence of the sample solution in the DPPH solution, A_2 was the absorbance of the blank control solution and A_3 was the absorbance of the sample solution without DPPH.

2.6.2. Measurement of Hydroxyl Radical Scavenging Activity

The hydroxyl radical scavenging activity was analyzed according to a reported method [37]. 2.0 mL of each sample solution was mixed with 2.0 mL of 6 mM $FeSO_4$, 2.0 mL of 6 mM salicylic acid and 2.0 mL of 6 mM H_2O_2, and then the mixture was incubated at 37 °C for 30 min. The absorbance was measured at 510 nm and the result was determined using the following equation:

$$\text{Hydroxyl radical scavenging activity } (\%) = [A_3 - (A_1 - A_2)]/A_3 \times 100\% \tag{3}$$

where A_1 was the absorbance of the sample solution, A_2 was the absorbance of the sample solution without H_2O_2 and A_3 was the absorbance of the control solution.

2.6.3. Measurement of Superoxide Anion Scavenging Activity

The superoxide anion scavenging activity was carried out using the method described previously [38]. 1.0 mL of the sample solution was added to 1.8 mL of 0.05 M Tris-HCl buffer (pH = 8.2), and then 100 μL of 0.01 M pyrogallic acid was added to the mixture. The absorbance was measured at 310 nm and the superoxide anion scavenging activity was calculated using the following equation:

$$\text{Superoxide anion scavenging activity } (\%) = (A_1 - A_2)/A_1 \times 100\% \tag{4}$$

where A_1 was the absorbance of the control solution and A_2 was the absorbance of the sample solution.

2.7. Statistical Analysis

The data were reported as the means ± SD (standard deviation) of three independent replicate experiments (n = 3). The statistics significance was evaluated using the *t*-test by the SPSS statistics 19.0 software (SPSS Inc., Chicago, IL, USA) and $P < 0.05$ was taken as significant.

3. Results and Discussion

3.1. Preparation of Lycopene/β-CD Inclusion Complexes

It has been reported that there are several different methods to synthesize CD inclusion complexes, such as freeze-drying, sealed-heating, ball-milling, solvent evaporation, spray drying, co-precipitation, neutralization and kneading [31]. In this study, the co-precipitation method was used to prepare the lycopene/β-CD inclusion complexes. A yield of the inclusion complexes was 83.0% by this way with an entrapment efficiency (the amount of lycopene in the inclusion complexes over the initial mass of lycopene used) of 71.8%.

Some common methods such as freeze-drying, precipitation and kneading were used for the preparation of the inclusion complex. The kneading method often showed a lower entrapment efficiency than freeze-drying and precipitation methods, these differences could be associated with the losses of active compounds during the kneading method which was carried out in an open container at room temperature. The high entrapment efficiencies were obtained by freeze-drying and precipitation methods, but the freeze-drying method is costly and time-consuming [39]. So, the precipitation methods were most used to prepare the inclusion complex due to its simple and low-cost characteristics. Some researcher also studied the difference of the entrapment efficiency between the different process such as magnetic stirring and ultrasonic in the precipitation method. For example, Gomes et al. (2014) prepared inclusion complexes of red bell pepper pigments with β-cyclodextrin using two different procedures (i.e., magnetic stirring and ultrasonic homogenisation), the results showed that the ultrasonic homogenisation procedure provided a higher yield compared to the magnetic stirring process while the evaluation of the inclusion efficiencies showed no significant difference between the two procedures [40]. Our research found that the yields and entrapment efficiencies in ultrasonic homogenisation procedure (data not shown) were both lower than that in the magnetic stirring process.

3.2. Characterization of Inclusion Complexes

3.2.1. Scanning Electron Microscopy (SEM)

The scanning electron micrographs of lycopene, β-CD, lycopene/β-CD inclusion complexes and their physical mixtures are shown in Figure 1. The pure lycopene (Figure 1a) was irregular-shaped particles with block structure, while β-CD displayed ellipsoidal form with different sizes (Figure 1b). The physical mixtures (Figure 1c) presented some similarities with the free molecules of lycopene and β-CD and showed both common structure characteristics. However, the lycopene/β-CD inclusion complexes (Figure 1d) showed a compact structure and were different from lycopene and β-CD in sizes and shapes, conforming the formation of inclusion complexes.

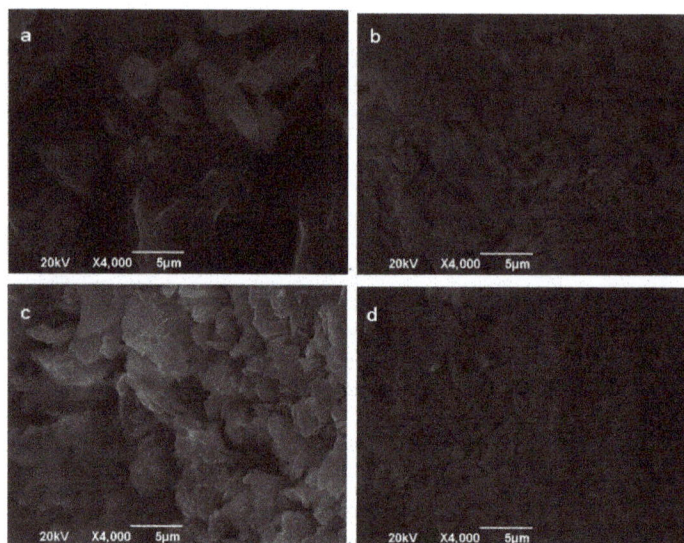

Figure 1. SEM of lycopene (**a**), β-CD (**b**), their physical mixtures (**c**) and inclusion complexes (**d**).

3.2.2. Microscopic Observation

The microscopic pictures of lycopene, β-CD, lycopene/β-CD inclusion complexes and their physical mixtures are shown in Figure 2. The pure lycopene (Figure 2a) was red acicular particles and β-CD (Figure 2b) displayed regular-shaped mesh structures. In the micrograph of the physical mixtures (Figure 2c), the lycopene and β-CD were found to exist overlap side by side. Nevertheless, whether the inclusion complexes appeared as an aqueous solution (Figure 2d) or as a crystal structure (Figure 2e), the lycopene was observed to be encapsulated by β-CD, which indicated the formation of the inclusion complexes between lycopene and β-CD.

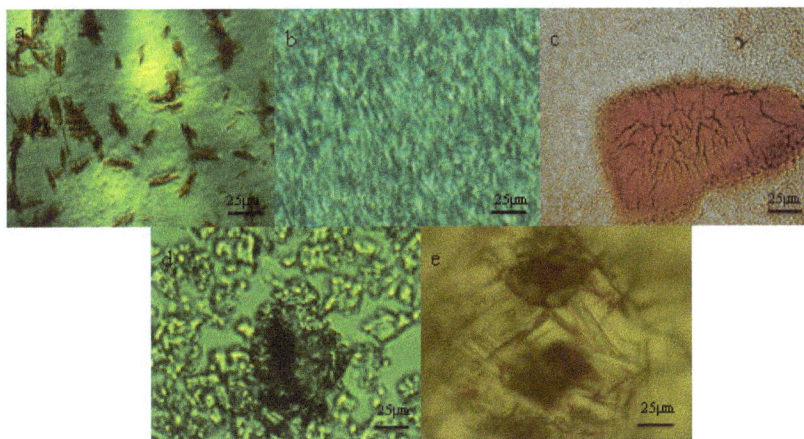

Figure 2. Micrograph of lycopene (**a**), β-CD (**b**), their physical mixtures (**c**) and inclusion complexes (aqueous solution (**d**), crystal structure (**e**)).

3.2.3. UV Analysis

As shown in Figure 3a, the characteristic absorption peaks of lycopene (dissolved in acetone) were found in 473 nm and 503 nm, while in Figure 3b, the UV absorbance of β-CD (dissolved in water) was extremely low and the characteristic absorption peak was found in 288 nm. When the inclusion complexes were extracted with acetone, we observed that its UV absorption (Figure 3c) was the same as that of lycopene, which showed that the inclusion complexes contained lycopene. Then in Figure 3d, the characteristic absorption peaks of the aqueous solution of the inclusion complexes was similar to that of the β-CD, but it was different from the UV absorption of lycopene, which indicated that the UV absorption of lycopene was covered by the external β-CD, thereby it showed the UV absorption similar to that of the external β-CD. These results indicated the formation of the inclusion complexes.

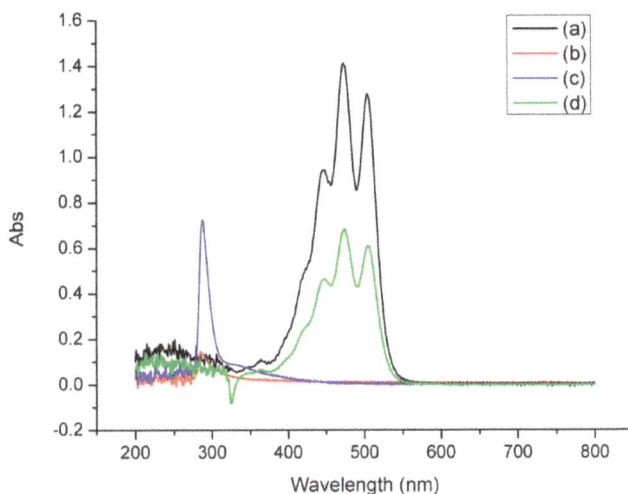

Figure 3. UV spectra of lycopene (**a**), β-CD (**b**) and inclusion complexes (aqueous solution (**c**), extracted with acetone (**d**)).

3.2.4. High Performance Liquid Chromatography Analysis

From Figure S1 in Supplementary Materials, the retention time of lycopene was observed at 18.911 min. However, the peak of β-CD (Figure S1b) was not observed, this might be because β-CD had no absorbance at 474 nm. At the same time, the chromatogram of lycopene/β-CD (Figure S1c in Supplementary Materials) showed no chromatographic peak at the same position as lycopene, but was similar to the result of β-CD, which indicated that the lycopene was embedded into the cavity of the β-CD, so the chromatographic retention time of lycopene/β-CD was different from that of lycopene. What's more, the chromatograms of β-CD (Figure S1d in Supplementary Materials) and lycopene/β-CD (Figure S1e in Supplementary Materials) were similar when the detection wavelength was 285 nm, which illustrated furtherly that the lycopene was embedded by β-CD, therefore the inclusion complexes exhibited more similar chromatographic characteristics to β-CD.

3.2.5. Differential Scanning Calorimetry (DSC)

Figure S2 in Supplementary Materials shows the DSC thermograms of lycopene, β-CD, their physical mixtures and the inclusion complexes. The thermogram of lycopene (Figure S2a in Supplementary Materials) showed two major peaks: One peak around 44 °C, probably because of loss of water, another peak at about 162.5 °C, likely due to its melting point. In the thermogram of β-CD (Figure S2b in Supplementary Materials), a single endothermic peak was observed at about 160 °C which was associated with its dehydration. For the physical mixtures (Figure S2c in Supplementary Materials), the thermogram was just the simple superposition of endothermic peaks of free species. However, the DSC curve of the inclusion complexes (Figure S2d in Supplementary Materials) showed different features of free molecules and the physical mixtures, indicating that there was probable interaction between the lycopene and β-CD. These results evidenced that the lycopene was embedded into the cavity of the β-CD.

3.3. Phase-Solubility Study

The phase-solubility diagram of the lycopene in β-CD solution is shown in Figure 4. As could be seen from Figure 4, the lycopene solubility increased linearly with the increasing concentration of β-CD. This diagram exhibited a classic A_L type model and the inclusion complexes were formed at a stoicheiometry of 1:1 [35,41]. This result indicated that the β-CD had a solubilizing effect on lycopene, showed that β-CD could entrap lycopene.

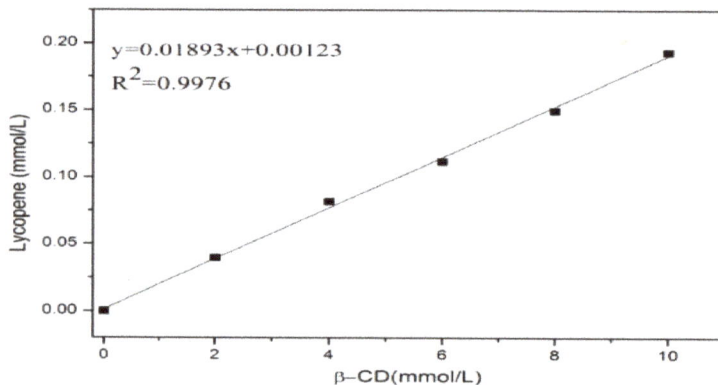

$$y=0.01893x+0.00123$$
$$R^2=0.9976$$

Figure 4. Phase-solubility diagram of lycopene in the presence of β-CD.

3.4. Stability Experiments

The thermal and photo stabilities of free and complexed lycopene were investigated. As shown in Figure 5a, the thermal degradation of lycopene was quicker than that of the inclusion complexes. The results indicated that the inclusion complexes could improve the thermal stability of lycopene in the solution.

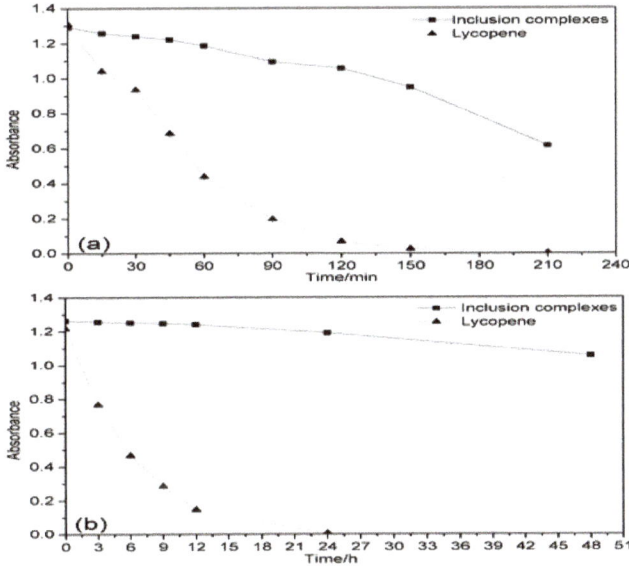

Figure 5. The effect of thermal (**a**) and illumination (**b**) on the stability of the lycopene and inclusion complexes.

From Figure 5b, the lycopene solution was vulnerable under light condition. After 24 h, the absorbance of lycopene in the solution had not been detected. However, when lycopene was encapsulated with β-CD, the absorbance of the inclusion complexes slowly reduced over time, the stability of lycopene/β-CD was significantly improved compared to the original lycopene. This indicated that the inclusion complexes could obviously enhance the photostability of lycopene.

These results show that when lycopene and β-CD were formed into the inclusion complex, the heat and illumination stability of lycopene were both greatly enhanced. Complexation was helpful to improve the stability of lycopene under storage condition. These results were also agreed with the former report [42]. The stability study of the encapsulated lycopene at room temperature was carried out by Blanch et al. (2007) and they found that no variation of the spectral signals shown by the β-CD/lycopene complex after six months. Consequently, they concluded that the lycopene complex with β-CD remains stable at least during half a year [29]. However, the detailed data were not shown. The stability studies of the encapsulated lycopene under thermal and illumination conditions were also not studied.

3.5. Antioxidant Activity

The antioxidant activities of lycopene and lycopene/β-CD complexes were measured using the stable free radical and the results are shown in Figure 6. As shown in Figure 6a, when the concentration of the lycopene and inclusion complexes were 5 μg/mL–30 μg/mL, the DPPH scavenging activity of the lycopene/β-CD complexes was higher than those of the lycopene. These results indicated that the antioxidant activity of lycopene was increased by the formation of the inclusion complexes with β-CD. However, from Figure 6b,c, the hydroxyl and superoxide anion radicals scavenging activities of the lycopene inclusion complexes were not increased significantly in the concentration range of 5 µg/mL–40 µg/mL. A similar study about the antioxidant activity of the inclusion complex of astaxanthin with hydroxypropyl-β-cyclodextrin was carried out by Yuan et al. (2013). Interestingly, the results show that the DPPH radical scavenging activity of the native lycopene were lower than complex with hydroxypropyl-β-cyclodextrin at low concentration while the hydroxyl radical scavenging activities of the complex was a little lower than that of the native lycopene [43]. These results agreed with our research and the reasons are worth further study.

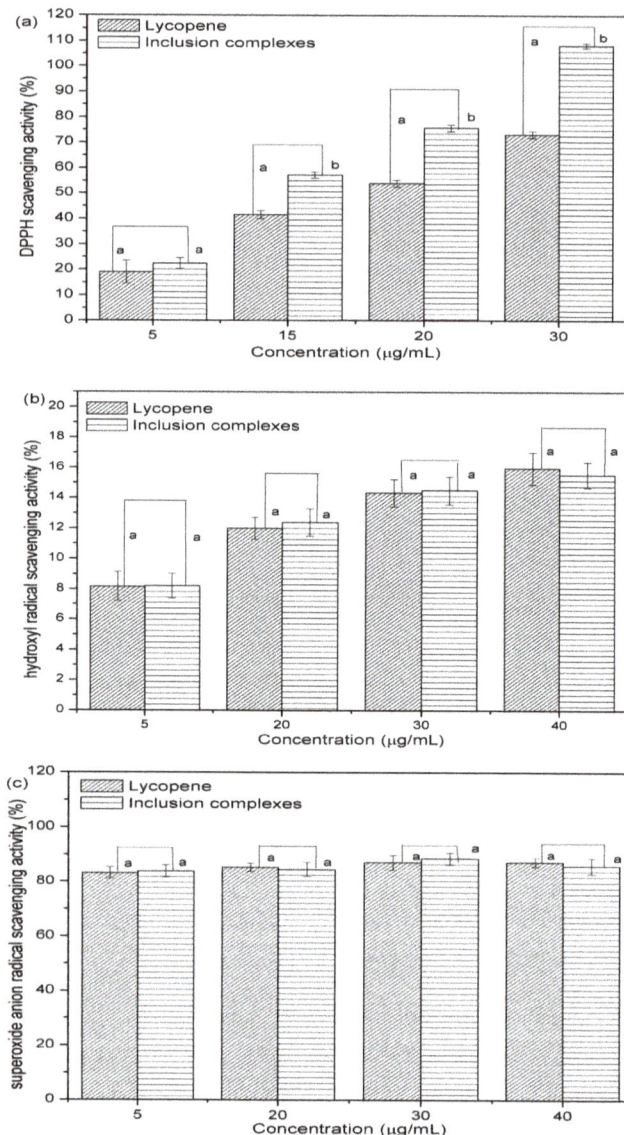

Figure 6. Antioxidant activity of the lycopene and inclusion complexes: (**a**) DPPH scavenging activity; (**b**) hydroxyl scavenging activity; (**c**) superoxide anion radical scavenging activity.

4. Conclusions

In this study, the inclusion complexes of lycopene with β-CD was successfully prepared by the co-precipitation method, the results of the yield and the entrapment efficiency of the inclusion complexes were 83.0% and 71.8%, respectively. Moreover, the results of SEM, Microscopic observation, UV, HPLC and DSC analyses confirmed clearly that the lycopene was embedded into the cavity of the β-CD and the formation of the inclusion complexes. Furthermore, the results of the phase-solubility study demonstrated that the β-CD had a solubilizing effect on lycopene and the stoichiometry of the inclusion complexes was 1:1. Furthermore, the thermal and photostability as well as the antioxidant activity of lycopene were all significantly increased by the formation of the lycopene/β-CD inclusion complexes.

Supplementary Materials: The following are available online at http://www.mdpi.com/2076-3921/8/8/314/s1, Figure S1: HPLC of lycopene (a), β-CD (b) and inclusion complexes (c) (Detection wavelength: 474nm); β-CD (d) and inclusion complexes (e) (Detection wavelength: 285nm), Figure S2: DSC thermograms of lycopene (a), β-CD (b), their physical mixtures (c) and inclusion complexes (d).

Author Contributions: Project Administration, H.W.; Investigation, S.W. (Shaofeng Wang); Original Draft Preparation, H.Z.; Supervision, S.W. (Suilou Wang); Funding Acquisition, J.X.

Funding: This research and the APC was funded by the Beijing Advanced Innovation Center for Food Nutrition and Human Health (grant number 20181009).

Acknowledgments: This work was supported by the open project funds of the Beijing Advanced Innovation Center for Food Nutrition and Human Health (grant number 20181009).

Conflicts of Interest: The authors declare that there is no conflict of interest.

References

1. Hernández-Almanza, A.; Montanez, J.; Martinez, G.; Aguilar-Jimenez, A.; Contreras-Esquivel, J.C.; Aguilar, C.N. Lycopene: Progress in microbial production. *Trends Food Sci. Technol.* **2016**, *56*, 142–148. [CrossRef]
2. Rao, A.V.; Agarwal, S. Role of lycopene as antioxidant carotenoid in the prevention of chronic diseases: A review. *Nutr. Res.* **1999**, *19*, 305–323. [CrossRef]
3. Srivastava, S.; Srivastava, A.K. Lycopene; chemistry, biosynthesis, metabolism and degradation under various abiotic parameters. *J. Food Sci. Technol.* **2015**, *52*, 41–53. [CrossRef]
4. Kim, C.H.; Park, M.K.; Kim, S.K.; Cho, Y.H. Antioxidant capacity and anti-inflammatory activity of lycopene in watermelon. *Int. J. Food Sci. Technol.* **2014**, *49*, 2083–2091. [CrossRef]
5. Hazewindus, M.; Haenen, G.R.M.M.; Weseler, A.R.; Bast, A. The anti-inflammatory effect of lycopene complements the antioxidant action of ascorbic acid and α-tocopherol. *Food Chem.* **2012**, *132*, 954–958. [CrossRef]
6. Rafi, M.M.; Kanakasabai, S.; Reyes, M.D.; Bright, J.J. Lycopene modulates growth and survival associated genes in prostate cancer. *J. Nutr. Biochem.* **2013**, *24*, 1724–1734. [CrossRef]
7. Sahin, K.; Cross, B.; Sahin, N.; Ciccone, K.; Suleiman, S.; Osunkoya, A.O.; Master, V.; Harris, W.; Carthon, B.; Mohammad, R.; et al. Lycopene in the prevention of renal cell cancer in the TSC2 mutant Eker rat model. *Arch. Biochem. Biophys.* **2015**, *572*, 36–39. [CrossRef] [PubMed]
8. Costa-Rodrigues, J.; Pinho, O.; Monteiro, P.R.R. Can lycopene be considered an effective protection against cardiovascular disease? *Food Chem.* **2018**, *245*, 1148–1153. [CrossRef]
9. Müller, L.; Caris-Veyrat, C.; Lowe, G.; Böhm, V. Lycopene and Its Antioxidant Role in the Prevention of Cardiovascular Diseases-A Critical Review. *Crit. Rev. Food Sci. Nutr.* **2016**, *56*, 1868–1879. [CrossRef]
10. Strati, I.F.; Oreopoulou, V. Recovery of carotenoids from tomato processing by-products-A review. *Food Res. Int.* **2014**, *65*, 311–321. [CrossRef]
11. Cadoni, E.; Rita De Giorgi, M.; Medda, E.; Poma, G. Supercritical CO_2 extraction of lycopene and β-carotene from ripe tomatoes. *Dyes Pigment.* **1999**, *44*, 27–32. [CrossRef]
12. Ciriminna, R.; Fidalgo, A.; Meneguzzo, F.; Ilharco, L.M.; Pagliaro, M. Lycopene: Emerging Production Methods and Applications of a Valued Carotenoid. *ACS Sustain. Chem. Eng.* **2016**, *4*, 643–650. [CrossRef]
13. Chen, J.; Shi, J.; Xue, S.J.; Ma, Y. Comparison of lycopene stability in water-and oil-based food model systems under thermal-and light-irradiation treatments. *LWT-Food Sci. Technol.* **2009**, *42*, 740–747. [CrossRef]

14. Shi, J.; Dai, Y.; Kakuda, Y.; Mittal, G.; Xue, S.J. Effect of heating and exposure to light on the stability of lycopene in tomato purée. *Food Control* **2008**, *19*, 514–520. [CrossRef]

15. Lee, M.T.; Chen, B.H. Stability of lycopene during heating and illumination in a model system. *Food Chem.* **2002**, *78*, 425–432. [CrossRef]

16. Pérez-Masiá, R.; Lagaron, J.M.; Lopez-Rubio, A. Morphology and Stability of Edible Lycopene-Containing Micro-and Nanocapsules Produced Through Electrospraying and Spray Drying. *Food Bioprocess Technol.* **2015**, *8*, 459–470. [CrossRef]

17. Rocha-Selmi, G.A.; Favaro-Trindade, C.S.; Grosso, C.R.F. Morphology, Stability, and Application of Lycopene Microcapsules Produced by Complex Coacervation. *J. Chem.* **2013**, *2013*, 1–7. [CrossRef]

18. Cheng, Y.S.; Lu, P.M.; Huang, C.Y.; Wu, J.J. Encapsulation of lycopene with lecithin and α-tocopherol by supercritical antisolvent process for stability enhancement. *J. Supercrit. Fluids* **2017**, *130*, 246–252. [CrossRef]

19. Dos Santos, P.P.; Paese, K.; Guterres, S.S.; Pohlmann, A.R.; Jablonski, A.; Flôres, S.H.; de Oliveira Rios, A. Stability study of lycopene-loaded lipid-core nanocapsules under temperature and photosensitization. *LWT-Food Sci. Technol.* **2016**, *71*, 190–195. [CrossRef]

20. Bou, R.; Boon, C.; Kweku, A.; Hidalgo, D.; Decker, E.A. Effect of different antioxidants on lycopene degradation in oil-in-water emulsions. *Eur. J. Lipid Sci. Technol.* **2011**, *113*, 724–729. [CrossRef]

21. Astray, G.; Gonzalez-Barreiro, C.; Mejuto, J.C.; Rial-Otero, R.; Simal-Gándara, J. A review on the use of cyclodextrins in foods. *Food Hydrocoll.* **2009**, *23*, 1631–1640. [CrossRef]

22. Szente, L.; Szejtli, J. Cyclodextrins as food ingredients. *Trends Food Sci. Technol.* **2004**, *15*, 137–142. [CrossRef]

23. Mura, P. Analytical techniques for characterization of cyclodextrin complexes in aqueous solution: A review. *J. Pharm. Biomed. Anal.* **2014**, *101*, 238–250. [CrossRef]

24. Abarca, R.L.; Rodríguez, F.J.; Guarda, A.; Galotto, M.J.; Bruna, J.E. Characterization of beta-cyclodextrin inclusion complexes containing an essential oil component. *Food Chem.* **2015**, *196*, 968–975. [CrossRef] [PubMed]

25. Dos Santos, C.; Buera, P.; Mazzobre, F. Novel trends in cyclodextrins encapsulation. Applications in food science. *Curr. Opin. Food Sci.* **2017**, *16*, 106–113. [CrossRef]

26. Durante, M.; Lenucci, M.S.; Marrese, P.P.; Rizzi, V.; De Caroli, M.; Piro, G.; Fini, P.; Russo, G.L.; Mita, G. α-Cyclodextrin encapsulation of supercritical CO_2 extracted oleoresins from different plant matrices: A stability study. *Food Chem.* **2016**, *199*, 684–693. [CrossRef] [PubMed]

27. Bruno, A.; Durante, M.; Marrese, P.P.; Migoni, D.; Laus, M.N.; Pace, E.; Pastore, D.; Mita, G.; Piro, G.; Lenucci, M.S. Shades of red: Comparative study on supercritical CO_2 extraction of lycopene-rich oleoresins from gac, tomato and watermelon fruits and effect of the α-cyclodextrin clathrated extracts on cultured lung adenocarcinoma cells' viability. *J. Food Compos. Anal.* **2018**, *65*, 23–32. [CrossRef]

28. Durante, M.; Lenucci, M.S.; Gazza, L.; Taddei, F.; Nocente, F.; De Benedetto, G.E.; De Caroli, M.; Piro, G.; Mita, G. Bioactive composition and sensory evaluation of innovative spaghetti supplemented with free or α-cyclodextrin chlatrated pumpkin oil extracted by supercritical CO_2. *Food Chem.* **2019**, *294*, 112–122. [CrossRef]

29. Blanch, G.P.; del Castillo, M.L.; del Mar Caja, M.; Pérez-Méndez, M.; Sánchez-Cortés, S. Stabilization of all-trans-lycopene from tomato by encapsulation using cyclodextrins. *Food Chem.* **2007**, *105*, 1335–1341. [CrossRef]

30. Del Valle, E.M.M. Cyclodextrins and their uses: A review. *Process Biochem.* **2004**, *39*, 1033–1046. [CrossRef]

31. Liu, B.; Zhu, X.; Zeng, J.; Zhao, J. Preparation and physicochemical characterization of the supramolecular inclusion complex of naringin dihydrochalcone and hydroxypropyl-β-cyclodextrin. *Food Res. Int.* **2013**, *54*, 691–696. [CrossRef]

32. Wang, J.; Cao, Y.; Sun, B.; Wang, C. Physicochemical and release characterisation of garlic oil-β-cyclodextrin inclusion complexes. *Food Chem.* **2011**, *127*, 1680–1685. [CrossRef]

33. Zhu, G.; Xiao, Z.; Zhu, G. Preparation, characterization and the release kinetics of mentha-8-thiol-3-one-β-cyclodextrin inclusion complex. *Polym. Bull.* **2017**, *74*, 2263–2275. [CrossRef]

34. Zhu, G.; Xiao, Z.; Zhou, R.; Zhu, Y. Study of production and pyrolysis characteristics of sweet orange flavor-β-cyclodextrin inclusion complex. *Carbohydr. Polym.* **2014**, *105*, 75–80. [CrossRef] [PubMed]

35. Xu, J.; Zhang, Y.; Li, X.; Zheng, Y. Inclusion complex of nateglinide with sulfobutyl ether β-cyclodextrin: Preparation, characterization and water solubility. *J. Mol. Struct.* **2017**, *1141*, 328–334. [CrossRef]

36. Mishra, K.; Ojha, H.; Chaudhury, N.K. Estimation of antiradical properties of antioxidants using DPPH-assay: A critical review and results. *Food Chem.* **2012**, *130*, 1036–1043. [CrossRef]

37. Chen, R.; Liu, Z.; Zhao, J.; Chen, R.; Meng, F.; Zhang, M.; Ge, W. Antioxidant and immunobiological activity of water-soluble polysaccharide fractions purified from *Acanthopanax senticosu*. *Food Chem.* **2011**, *127*, 434–440. [CrossRef] [PubMed]

38. Zhao, Z.Y.; Huangfu, L.T.; Dong, L.L.; Liu, S.L. Functional groups and antioxidant activities of polysaccharides from five categories of tea. *Ind. Crops Prod.* **2014**, *58*, 31–35. [CrossRef]

39. Tao, F.; Hill, L.E.; Peng, Y.; Gomes, C.L. Synthesis and characterization of β-cyclodextrin inclusion complexes of thymol and thyme oil for antimicrobial delivery applications. *LWT-Food Sci Technol.* **2014**, *59*, 247–255. [CrossRef]

40. Gomes, L.M.M.; Petito, N.; Costa, V.G.; Falcão, D.Q.; De Lima Araújo, K.G. Inclusion complexes of red bell pepper pigments with β-cyclodextrin: Preparation, characterisation and application as natural colorant in yogurt. *Food Chem.* **2014**, *148*, 428–436. [CrossRef]

41. Yang, L.J.; Wang, S.H.; Zhou, S.Y.; Zhao, F.; Chang, Q.; Li, M.Y.; Chen, W.; Yang, X.D. Supramolecular system of podophyllotoxin and hydroxypropyl-β-cyclodextrin: Characterization, inclusion mode, docking calculation, solubilization, stability and cytotoxic activity. *Mater. Sci. Eng. C* **2017**, *76*, 1136–1145. [CrossRef] [PubMed]

42. Chen, X.; Chen, R.; Guo, Z.; Li, C.; Li, P. The preparation and stability of the inclusion complex of astaxanthin with β-cyclodextrin. *Food Chem.* **2007**, *101*, 1580–1584. [CrossRef]

43. Yuan, C.; Du, L.; Jin, Z.; Xu, X. Storage stability and antioxidant activity of complex of astaxanthin with hydroxypropyl-β-cyclodextrin. *Carbohydr. Polym.* **2013**, *91*, 385–389. [CrossRef] [PubMed]

MDPI

St. Alban-Anlage 66

4052 Basel

Switzerland

Tel. +41 61 683 77 34

Fax +41 61 302 89 18

www.mdpi.com

Antioxidants Editorial Office

E-mail: antioxidants@mdpi.com

www.mdpi.com/journal/antioxidants

www.ingramcontent.com/pod-product-compliance
Lightning Source LLC
Chambersburg PA
CBHW041216220326
41597CB00033BA/5991